Climate Change
DEBUNKED

There Is No Impending Catastrophic Doom

Dr. Robert D. Oppenheimer, M.D.

Copyright © 2010 by Robert D. Oppenheimer

All rights reserved. No part of this publication may be reproduced or transmitted in any form or by any means electronic or mechanical, including photocopy, recording, or any information storage and retrieval system now known or to be invented, without permission in writing from the author, except by a reviewer who wishes to quote brief passages in connection with a review written for inclusion in a magazine, newspaper, or broadcast.

ISBN-13: 978-1-453-71508-6

To my beautiful Sasha, for being understanding, for putting up with my spending all day on the computer, and for so many other things.

Contents

Part 1   **Introduction**
- 1   A Brief History 1
- 2   The Consensus On Global Warming 11

Part 2   **Was Our Recent Warm Period Unique?**
- 3   Previous Rates Of Warming 31
- 4   The Medieval Warm Period 39
- 5   The Solar Connection 65
- 6   Recent Cooling 75
- 7   Ice Melts 89

Part 3   **Scientific Dishonesty**
- 8   The Himalayan Glacier Lie 111
- 9   The Quality Of U.S. Temperature Data 125
- 10   More On Bad Temperature Data 151
- 11   Hide The Decline 161
- 12   Bullying, Hiding, And The Money 179
- 13   References 213

Chapter 1

A Brief History

"As with the World Health Organization's claim that 150,000 people are killed by the effects of climate change, each and every single one of these deaths would have been avoided had there been the level of development in those regions as there is in the West. It cannot be argued, therefore, that climate change is responsible for those deaths. Poverty was responsible for them."

- Ben Pile and Stewart Blackman

A miracle happened. On November 17th, 2009, an anonymous Internet user using the name FOIA [Freedom of Information Act] posted the following message on a global warming discussion website:

"We feel that climate science is, in the current situation, too important to be kept under wraps.

We hereby release a random selection of correspondence, code, and documents. Hopefully it will give some insight into the science and the people behind it."

Linked to their post was a 61-megabyte archive that contained thousands of documents and over 1,000 emails to and from the Climatic Research Unit (CRU) at East Anglia University in Norwich, England.

This archive turned out to be a damning tomb of evidence for the scientists working with the CRU. Evidence of data manipulation, data misrepresentation, tax evasion, bullying, and all sorts of other unscientific behavior was found. From this point forward, the global warming hypothesis will never be the same.

This leak of information has become known as ClimateGate. The scientists involved have become known as the ClimateGate scientists.

It is currently unknown who leaked the files, but there are several good guesses.

Some suggest that an unhappy computer programmer leaked the files, while others suggest it was one of the climate scientists who grew tired of being bullied by the other scientists. But that is all speculation.

These documents give us insight into the "behind the scenes" conversations of climate scientists. These emails will be quoted, when appropriate, throughout this book. A lot of the emails contained spelling errors and abbreviations. I have corrected and fixed these as appropriate in order to make the emails as lucid as possible. Each email is annotated with a web address to access the original email, which you are welcome to look up if you feel I have misrepresented any email in any way.

When reading these emails, you will see certain names repeatedly. I will now give a brief introduction to each of the key players in the ClimateGate scandal.

Michael Mann is the lead ClimateGate conspirator in the United States. He is a geologist / geophysicist working out of Pennsylvania State University and is one of the most controversial figures we will see throughout the book. His specialty is paleoclimatology (the study of past climate) and he is notorious for his "hockey stick graph" which we will learn about later. At the time of this writing he is being investigated by his employer. Other organizations are waiting in line to investigate him once Penn State is through.

Phil Jones is the lead ClimateGate conspirator in the United Kingdom. He was the head climate scientist at the Climatic Research Unit at East Anglia University in Norwich, England, until ClimateGate broke, at which point he stepped down. He has worked extensively with Michael Mann and Keith Briffa and a lot of their papers are the basis for much of climate science.

Tom Wigley is an older conspirator who becomes increasingly worried about the unfolding scandal. He is an interesting character to follow throughout the emails. He is the former director at the Climatic Research Unit and now works for the University Corporation for Atmospheric Research.

Keith Briffa is an older conspirator whose views are sometimes more in line with reality and sometimes quite bizarre. He appears to not care much for Michael Mann and his methodologies. His specialty is paleoclimatology, primarily using dendrochronology (using tree rings to learn about past climate.) He has more or less gone missing since ClimateGate broke.

Lastly, the U.K. Parliament is starting an investigation of the ClimateGate scandal and is looking into the poor quality of data issues from the Climatic Research Unit at East Anglia University. The Parliament's Science and Technology Committee recently announced an inquiry and are asking people to write to them to help them understand the implications of the emails.

As of this writing, ClimateGate is still relatively new and the implications of what has been revealed have not yet been fully realized. It will take years to examine everything that ClimateGate has revealed.

GLOBAL WARMING

The global warming hypothesis is the idea that in the 4,500,000,000 years of earth's existence the warming period since 1975 has occurred at a rate that is greater than could be expected by natural forces. That is, the global warming hypothesis states that our most recent warming period is both unusual and unprecedented. This hypothesis maintains that the cause for the accelerated rate of warming is human-produced greenhouse gases, namely carbon dioxide.

Indeed, carbon dioxide is a greenhouse gas, and it has the potential to warm the planet. But these greenhouse gases in our atmosphere do not behave like an actual greenhouse[1]. A greenhouse works by acting as a barrier to prevent the transfer of heat (convection.) They are designed to let heat into the room, but not to let heat out of the room.

In contrast, our atmosphere, via the greenhouse effect, actually facilitates the transfer of heat away from the planet. Greenhouse gases do not trap the suns heat. The greenhouse effect works as follows.

The sun emits high energy, shortwave radiation. Our greenhouse gases are mostly transparent to this radiation, and this radiation hits earth. In response, the earth emits low energy, longwave radiation. The greenhouse gases are opaque to longwave radiation, and absorb it and transfer it through different ways. Some of this energy comes back to earth. When they transfer (or emit) energy, they do not emit it in the same wavelengths as they absorbed it. Therefore, emissions from the greenhouse gas carbon dioxide are not absorbed again by carbon dioxide.

This means that heat is not "trapped" in the atmosphere forever. This means the heating via greenhouse gases is not a runaway, self-amplifying process. Heat is only temporarily delayed until it returns back to space.

This allows us to see that the greenhouse gases do not act like a blanket around the earth, because they do not function as a real greenhouse. They merely delay the exit of heat, not trap it.

The greenhouse effect is essential for life to exist. Without it, the earth would be approximately -18.0' Celsius and the presence of the effect raises our temperatures to a much more habitable 33.0' Celsius.

The majority of the greenhouse effect is caused by water, in the form of water vapor and clouds. In the total atmosphere, water accounts for approximately 90% of the earth's greenhouse effect. The remaining 10% or so of the greenhouse effect comes from other sources, such as carbon dioxide, nitrous oxide, methane, ozone, and other trace gases. These are approximations, and as we learn more about the way the atmosphere works, we keep decreasing the force that carbon dioxide exerts on the greenhouse effect. In fact, the 10% estimate may be an order of magnitude too high. Time will tell.

In terms of quantity, carbon dioxide is a trace gas in the atmosphere. Approximately 0.038% of the atmosphere is made up of carbon dioxide. This is up from approximately 0.028% during the industrial revolution.

Clearly, we can see that carbon dioxide is a minor player in the greenhouse effect and that we have contributed relatively minor increases in carbon dioxide concentration in the atmosphere.

These facts run contrary to those who believe in the global warming hypothesis. They necessarily believe that carbon dioxide is the dominant greenhouse gas. They believe that the climate system is very sensitive to changes in carbon dioxide, and that even tiny increases of this gas in the atmosphere will lead to great changes in warming.

These predictions of water vapor vs. carbon dioxide were somewhat difficult to quantify until recently. A recent paper, Solomon et al. 2009[2], found that increases in water vapor in the stratosphere lead to increased rates of warming, and they found that decreased levels of water vapor in the stratosphere lead to decreased rates of warming. These findings are consistent with the fact that stratospheric water vapor represents an important driver of global surface climate change.

Contrast that to the fact that carbon dioxide levels have been slowly rising since 1850, and in that time period we have had multiple periods of warming separated by vast (25 – 40 years) periods of cooling. In fact, in a later chapter, we will show that there has been global cooling 1998 or 2001 despite rising carbon dioxide levels. This implies a negative correlation between the greenhouse gas and global mean temperatures.

Catastrophic anthropogenic global warming is the idea that human created (anthropogenic) greenhouse gases are contributing to the rapid destruction of the earth. People who believe that we are experiencing catastrophic anthropogenic global warming are called alarmists. Those who reject these ideas are termed skeptics.

Alarmists use fallacious arguments in an attempt to appeal to your emotions. They make predictions that paint horrible pictures of sea levels rising and wiping out half of the world, rainforests dying, huge glaciers melting and about an increase in terrible, destructive weather. Alarmists blame global warming for everything from dying polar bears to the increased spread of HIV.

WHERE WE'RE AT NOW

The earth's climate is dynamic. It is constantly changing, and it is infinitely complicated. This is the reason that weather models cannot forecast several days of weather, and it's the same reason that climate models cannot forecast even three months of climate.

Approximately 1,000 years ago we experienced a period of very warm temperatures known as the Medieval Warm Period. The Medieval Warm Period (MWP) was approximately 1.0' Celsius warmer than now and lasted from about AD 800 – 1300 (more on the MWP in a later chapter.)

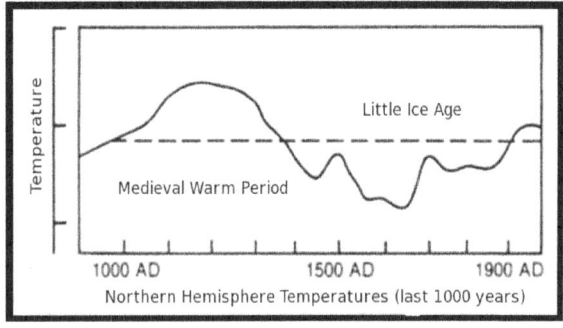

Figure 1 - 1,000 years of temperatures, including the Medieval Warm Period and the Little Ice Age

Following the Medieval Warm Period, the world fell into a Little Ice Age (LIA.) The Little Ice Age was a period of cooling that occurred from the 16th to the 19th centuries. It is generally agreed that there were three minima (periods of minimum warmth, or maximum coolness), beginning about 1650, about 1770, and 1850, each separated by intervals of slight warming.

It is the date 1850 that interests us. 1850 is known to be the last minima of the Little

Ice Age and it is also the start date that alarmists use when displaying graphs that show the warming of the earth. It is, of course, obvious that we will experience warming while coming out of the minima of a Little Ice Age. We would expect nothing else. Alarmists conveniently use this as a way to convince people that our global mean temperatures have been on a steep, unprecedented, upward rise for over 150 years.

Let me briefly express my views. I believe that human produced carbon dioxide has the potential to elevate global temperatures, but I think that we currently greatly overestimate its ability to do so. In other words, I believe anthropogenic emissions have marginally contributed to our most recent period of warming. I do not believe that our most recent period of warming is unique or unusual. I believe it is all part of a normal cycle.

I do believe that we rely too heavily on non-sustainable fuels and that we need to slowly switch over to renewable and sustainable energy sources. I support nuclear, solar, and wind power, but I do not think we should force conversion at cost of bankrupting the world's economy. We can provide economic incentives for making the switch, but we cannot impose taxes that will eliminate businesses that are unable to adapt to a non-problem.

In this book we will systematically destroy the idea of catastrophic anthropogenic global warming. We will show that our warming is not unique and that it is not occurring at a rate greater than expected by natural variation. We will show that the temperature record is poor and is not an accurate representation of global land temperatures, and we will show that the scientists pushing anthropogenic global warming do so by using deceit, bullying other scientists, and by forcing their data to fit their predetermined conclusions.

We will use selected ClimateGate emails to emphasize points and to show what is going on behind the scenes. This book is not meant to be an exhaustive demonstration of the ClimateGate emails. Thorough examinations of the emails have been performed by at least two parties. The first is by Stephen Mosher and Thomas W. Fuller, in their book "ClimateGate: The CRUtape Letters" and the second is by John P. Costella.[3]

This book will not contain a lot of fluff or other nonsense. We'll stick to the facts and look at the claims behind anthropogenic global warming claims. Thank you for reading.

Chapter 2

The Consensus On Global Warming

"When 20 per cent of earth scientists are anthropogenic [global warming] contrarians and a majority of oil geologists agree with them, it is easy to find skeptics. They do not, however, deserve to be called deniers. Some of them are Nobel Prize winners and they deserve to be heard"

- Calgary Herald Editorial

A topic that usually comes up in the great catastrophic anthropogenic global warming debate is whether or not there is a consensus among scientists that human produced greenhouse gases were the primary driving force behind the most recent warming episode we experienced (1975 - 1998).

The alarmist's argument:

"97% of climate scientists actively publishing climate papers endorse the position that human activity is a significant contributing factor in changing mean global temperatures." – 2009 Poll[4]

The skeptic's argument:

"The Petition Project features over 31,000 scientists signing the petition stating "there is no convincing scientific evidence that human release of carbon dioxide, methane, or other greenhouse gases will, in the foreseeable future, cause catastrophic heating of the Earth's atmosphere and disruption of the Earth's climate. Moreover, there is substantial scientific evidence that increases in atmospheric carbon dioxide may produce many beneficial effects upon the natural plant and animal environments of the Earth." – The Petition Project[5]

Let's look at both of these arguments.

THE ALARMIST'S ARGUMENT

We will start with the argument the alarmists use. They quote a 2009 poll in which 10,257 Earth scientists were invited to participate in an Internet-based poll that asked two questions:

1. When compared with pre-1800s levels, do you think that mean global temperatures have generally risen, fallen, or remained relatively constant?

2. Do you think human activity is a significant contributing factor in changing mean global temperatures?

Question 1 is absurd. I'm pretty sure basically everyone will agree that we have generally experienced warming since we came out of the Little Ice Age which ended in 1850 (although 3 climatologists in their poll disagreed - I'd love to know why.)

The answers to question 2 were the main results that were discussed in the poll. I strongly feel this question is very poorly worded. Does significant mean "an important contributing factor" or does it mean "the driving force"?

Does significant mean that it contributes to the majority of global warming, or does it mean that it contributes a minimal, but quantifiable amount? I think that carbon dioxide contributes to global mean temperatures, but I think we greatly exaggerate the amount. How would I answer that question?

Of the 10,257 people that were invited to participate, only 3,146 answered the poll. The vast majority of respondents were from the U.S., with the rest being from Canada and other countries. About 90% of respondents had PhDs, which is approximately 2,830. The rest had M.S. (masters) and B.S. (bachelors) degrees.

According to the poll, 97.4% of climate scientists (97.4% sounds impressive until you see that it means 75 people out of 77) who are actively publishing papers agree that carbon dioxide is a significant contributing factor to changing mean global temperature. For most alarmists, this is all they need to read. They see this result, and they run with it. They think that this means that 97.4% of climate scientists agree that we are on a crash course to hell, all the worlds ice is going to melt, the polar bears will die, everyone will get malaria and HIV, and we will lose most of our land because of rising sea levels.

But it's not that simple. Do these 75 (97.4%) climate scientists think that we are undergoing catastrophic warming as shown in the Intergovernmental Panel on Climate Change (IPCC) reports? Or do they believe that the warming is marginally greater than that of natural variation? Do they think that carbon dioxide is the sole factor contributing to warming? Or do they think that carbon dioxide has contributed to a minor (but quantifiable) amount of warming? Do they think that our current warming is unmatched in the last 1,000 years? Or do they think prior times were warmer than this? Do they think that dendrochronology is a junk science? (As Dr. Ed Cook implied[6] in the ClimateGate emails.) Or do they think that it is an appropriate way to look at paleoclimate? Do they think solar irradiance was a major input or not?

The problem is that the poll question is too simple. We cannot infer the answers to any of the above questions from the poll. It operates on the principle that if you ask a vague enough question then you can prove whatever you want. This poll is as clever as a carnival fortuneteller.

Let's look at what the climate scientist Keith Briffa thinks, and then we can further evaluate if there is really a scientific consensus. Keith Briffa is one of the climate scientists involved with the ClimateGate scandal. There is a lot of speculation[7] that he may be the whistleblower that leaked all of the ClimateGate emails and documents.

"Chris Folland, Phil Jones, Michael Mann,

Let me say that I don't mind what you put in the policy makers summary if there is a general consensus. However some general discussion would be valuable.

....

The multi proxy series (Mann et al. Jones et al) supposedly represent annual and summer seasons respectively, and both contain large proportions of tree-ring input. The latest tree-ring density curve (i.e. our data that have been processed to retain low frequency information) shows more similarity to the other two series- as do a number of other lower resolution data (Bradley et al, Peck et al ., and new Crowley series - see our recent Science piece) whether this represents 'TRUTH' however is a difficult problem.

I know Mike thinks his series is the 'best' and he might be right - but he may also be too dismissive of other data and possibly over confident in his (or should I say his use of other's). After all, the early (pre-instrumental) data are much less reliable as indicators of global temperature than is apparent in modern calibrations that include them and when we don't know the precise role of particular proxies in the earlier portions of reconstruction it remains problematic to assign genuine confidence limits at multi-decadal and longer timescales. I still contend that multiple regressions against the recent very trendy global mean series is potentially dangerous.

...

I do believe that it should not be taken as read that Mike's series (or Jones' et al. for that matter) is THE CORRECT ONE.

...

There is still a potential problem with non-linear responses in the very recent period of some biological proxies (or perhaps a fertilization through high CO2 or nitrate input).

I know there is pressure to present a nice tidy story as regards 'apparent unprecedented warming in a thousand years or more in the proxy data' but in reality the situation is not quite so simple. We don't have a lot of proxies that come right up to date and those that do (at least a significant number of tree proxies) some unexpected changes in response that do not match the recent warming. I do not think it wise that this issue be ignored in the chapter.

For the record, I do believe that the proxy data do show unusually warm conditions in recent decades. I am not sure that this unusual warming is so clear in the summer responsive data. I believe that the recent warmth was probably matched about 1000 years ago. I do not believe that global mean annual temperatures have simply cooled progressively over thousands of years as Mike appears to and I contend that that there is strong evidence for major changes in climate over the Holocene (not Milankovich) that require explanation and that could represent part of the current or future background variability of our climate. I think the Venice meeting will be a good place to air these issues."

- Keith Briffa[8]

In that very telling email we see at least four things: 1. Dr. Briffa believes the Medieval Warm Period was as warm or warmer than it is today (which means our current warming is not unique, which means he believes natural variation is more than enough to account for our current warming), 2. He is under pressure to manipulate his data to present a nice and tidy story of global warming, 3. He does not agree with Dr. Michael Mann on which tree ring reconstructions are the best ones (note the lack of a consensus about certain data within the ClimateGate team), and 4. He believes that Mann has misrepresented the data previously (about it cooling progressively over a thousand years – we will learn all about this in the Medieval Warm Period chapter.)

How would Briffa have answered the poll question? I believe he would have said that he thinks carbon dioxide is a significant factor that contributes to changing global mean temperatures and therefore would be included in the 97.4%. But I don't think that he believes we are on a crash course to hell caused by global warming. So would it really be right for an alarmist to claim that he is part of a scientific consensus that believes in what they believe?

Next, we have to look at the poll's definition of a climate scientist, which is as follows: "respondents that listed climate science as their area of expertise and who also have published more than 50% of their recent peer-reviewed papers on the subject of climate change." How many of those did they have that agreed? 75. If we look at just the U.S. that is 1.5 climatologists per state! The "consensus" is just 75 climatologists!

It gets worse. At first glance we would be tempted to believe those 75 people are all climatologists, but we would be wrong. It is impossible to know who responded to the poll, but we can cast doubt on some of the responders' qualifications by looking at some of the biggest players in the climatology game.

Let's start with Rajendra Pachauri. He is the head of the Intergovernmental Panel on Climate Change (IPCC.) They compile data that presents a persuasive picture of global warming and use it to try to convince the world's legislators that we are experiencing catastrophic warming. By the poll's definition he would be considered a climatologist. But his PhD is in Railroad Engineering[9].

We can also look at Michael Mann from Penn State. He is the climate "scientist" that is infamous for publishing the hockey stick tree ring graphs that have been thoroughly debunked multiple times. He's notorious for using horrible program code to present his data in widely distorted ways (so poorly that random red noise can be fed into his programs and they produce the same), he has bullied journal editors into not accepting skeptical climatology papers, and he has corrupted the peer review process. By the above definition he would be considered a climatologist, but his PhD is in geology / geophysics.

Next we have Gavin Schmidt, who founded the CRU / ClimateGate propaganda website RealClimate, credited as a climatologist. But his PhD is in applied mathematics.

And we also have Edward Cook, who has a PhD in watershed management. You have to love what Cook thinks about his field of study (dendrochronology):

> "Without trying to prejudice this work, but also because of what I almost think I know to be the case, the results of this study will show that we can probably say a fair bit about less than 100 year extra-tropical Northern Hemisphere temperature variability (at least as far as we believe the proxy estimates), but honestly know fuck-all about what the greater than 100 year variability was like with any certainty (i.e. we know with certainty that we know fuck-all)."

– *Edward Cook*[10]

In no way am I downplaying applied mathematics, geology, watershed management, or railroad engineering, nor am I saying that they are unqualified to discuss climate change, but they are not technically climatologists and should not be counted in a poll as climatologists. I am merely pointing this out to demonstrate that the poll's definition of a climatologist is flawed.

This leads us to the next problem with the poll. There is no way to audit the poll's results. We can't go through their data and see the names of the people who claim to be climatologists. We can't verify their credentials. We don't know if they actually exist. We don't know what kinds of papers they've published, which journals they've been in, if any, or anything else.

Next, arguing that someone's views are the right ones because a "scientific consensus" agrees with them is a logical fallacy known as an appeal to authority, in which it is argued that a statement is correct because the statement is made by a person or source that is commonly regarded as authoritative.

Lastly, science does not work by democracy. A consensus does not make a theory more or less true. The data supporting the theory is what makes it more or less true. We will learn how the ClimateGate scientists used the false idea of a consensus agreement to their advantage.

So, we have shown that the poll's questions were rather vague, that their definition of climatologist was rather vague, that we can't audit who responded to the poll, and that there is pressure put on climatologists to act as if global warming is a real threat.

THE SKEPTIC'S ARGUMENT

The Petition Project is a petition that consists of (currently) 31,478 signatures all of which support this very concrete statement against the idea of anthropogenic global warming:

"There is no convincing scientific evidence that human release of carbon dioxide, methane, or other greenhouse gases will, in the foreseeable future, cause catastrophic heating of the Earth's atmosphere and disruption of the Earth's climate. Moreover, there is substantial scientific evidence that increases in atmospheric carbon dioxide may produce many beneficial effects upon the natural plant and animal environments of the Earth."

The Nongovernmental International Panel on Climate Change (NIPCC) report strengthens this statement.[11] The NIPCC report is an 880 page book that demonstrates overwhelming scientific support for the position that the warming of the twentieth century was moderate and not unprecedented, that its impact on human health and wildlife was positive, and that carbon dioxide probably is not the driving factor behind climate change. The report is available free for download from the NIPCC's website.

Of those 31,478 signatures from the petition, 9,029 are PhDs, 7,153 M.S.s, 2,585 M.D.s or D.V.M.s, and 12,711 B.S.s. These are staggering numbers. This petition has 10 times the number of people than the alarmist poll had. This petition has three times more PhDs than the alarmists' poll had.

In this petition, each signer was painstakingly verified to see if they 1) exist, and 2) had the degree they claimed. The M.D. and D.V.M. signers were also required to have earned a degree in one of the hard sciences. All petition signers were required to be U.S. citizens or U.S. legal residents; if the Petition had been opened to scientists worldwide, the numbers would have been far higher.

This petition is powerful because it backs a very specific statement, as quoted above. It is not vague, and it makes it very clear what the people who signed it believe. This is the opposite of the alarmist poll, because the poll was rather vague in asking what people felt about the anthropogenic contributions of greenhouse gases.

Also, this petition is powerful because people had to voluntarily go out of their way to sign it. They had to contact the Petition Project, fill out a form, and mail it in. For the alarmist poll, the pollsters contacted the population and they were able to (much more easily) fill out a form on the Internet without having to get up from their desk.

Some criticisms of the petition project are of the population that makes up the signatures. Critics say that some of the people who have signed are not qualified to give their opinion on climate science, but at the same time they would let a railroad engineer head the IPCC. Let's look at a breakdown of the educational backgrounds of the petition project signers and see why each educational background is important to understanding climate change.

The following is directly from the NIPCC report:

"1. Atmospheric, environmental, and Earth sciences includes 3,803 scientists trained in specialties directly related to the physical environment of the Earth and the past and current phenomena that affect that environment.

2. Computer and mathematical sciences includes 935 scientists trained in computer and mathematical methods. Since the human-caused global warming hypothesis rests entirely upon mathematical computer projections and not upon experimental observations, these sciences are especially important in evaluating this hypothesis.

3. Physics and aerospace sciences include 5,810 scientists trained in the fundamental physical and molecular properties of gases, liquids, and solids, which are essential to understanding the physical properties of the atmosphere and Earth.

4. Chemistry includes 4,818 scientists trained in the molecular interactions and behaviors of the substances of which the atmosphere and Earth are composed.

5. Biology and agriculture includes 2,964 scientists trained in the functional and environmental requirements of living things on the Earth.

6. Medicine includes 3,046 scientists trained in the functional and environmental requirements of human beings on the Earth.

7. Engineering and general science includes 10,102 scientists trained primarily in the many engineering specialties required to maintain modern civilization and the prosperity required for all human actions, including environmental programs."

The outline below gives a more detailed analysis of the signers' educations. The numbers in parenthesis are the number of signatures in that particular area of study. The subjects in bold are the broad categories, and beneath them are a breakdown of the categories.

Atmosphere, Earth, and Environment (3,803)

1. Atmosphere (578)

    a) Atmospheric Science (113)
    b) Climatology (39)
    c) Meteorology (341)
    d) Astronomy (59)
    e) Astrophysics (26)

2. Earth (2,240)
    a) Earth Science (94)
    b) Geochemistry (63)
    c) Geology (1,684)
    d) Geophysics (341)
    e) Geoscience (36)
    f) Hydrology (22)

3. Environment (985)
    a) Environmental Engineering (486)
    b) Environmental Science (253)
    c) Forestry (163)
    d) Oceanography (83)

Computers and Math (935)

1. Computer Science (242)

2. Math (693)
    a) Mathematics (581)
    b) Statistics (112)

Physics and Aerospace (5,810)

1. Physics (5,223)
    a) Physics (2,365)
    b) Nuclear Engineering (223)
    c) Mechanical Engineering (2,635)

2. Aerospace Engineering (587)

Chemistry (4,818)

1. Chemistry (3,126)

2. Chemical Engineering (1,692)

Biochemistry, Biology, and Agriculture (2,964)

1. Biochemistry (744)
    a) Biochemistry (676)
    b) Biophysics (68)

2. Biology (1,437)
    a) Biology (1,048)
    b) Ecology (76)
    c) Entomology (59)
    d) Zoology (149)
    e) Animal Science (105)

3. Agriculture (783)
    a) Agricultural Science (296)
    b) Agricultural Engineering (114)
    c) Plant Science (292)
    d) Food Science (81)

Medicine (3,046)

1. Medical Science (719)

2. Medicine (2,327)

General Engineering and General Science (10,102)

    1. General Engineering (9,833)

        a) Engineering (7,280)

        b) Electrical Engineering (2,169)

        c) Metallurgy (384)

    2. General Science (269)

Again, these numbers are staggering. Of the 9,029 signers, there were 3,803 signers that specialize in atmosphere, earth, and environmental science. Compare that to the 75 that the alarmist poll had.

EVEN MORE SCIENTISTS

On December 8th, 2009, an open letter to the United Nations Secretary-General Ban Ki Moon was drafted which outlined the current state of climate science. They highlighted the point that climate science is not "settled" and 141 scientists signed this letter.[12]

Specifically, this letter outlined ten items that the proponents of the global warming hypothesis have still failed to prove:

"1. Variations in global climate in the last hundred years are significantly outside the natural range experienced in previous centuries;

2. Humanity's emissions of carbon dioxide and other 'greenhouse gases' (GHG) are having a dangerous impact on global climate;

3. Computer-based models can meaningfully replicate the impact of all of the natural factors that may significantly influence climate;

4. Sea levels are rising dangerously at a rate that has accelerated with increasing human GHG emissions, thereby threatening small islands and coastal communities;

5. The incidence of malaria is increasing due to recent climate changes;

6. Human society and natural ecosystems cannot adapt to foreseeable climate change as they have done in the past;

7. Worldwide glacier retreat, and sea ice melting in Polar Regions, is unusual and related to increases in human GHG emissions;

8. Polar bears and other Arctic and Antarctic wildlife are unable to adapt to anticipated local climate change effects, independent of the causes of those changes;

9. Hurricanes, other tropical cyclones and associated extreme weather events are increasing in severity and frequency;

10. Data recorded by ground-based stations are a reliable indicator of surface temperature trends."

They further point out that it is not the responsibility of "climate realist" scientists (skeptics) to prove that dangerous human-caused climate change is not happening. Rather, it is those who propose that it is, and promote the allocation of massive investments to solve the supposed 'problem', who have the obligation to convincingly demonstrate that recent climate change is not of mostly natural origin and, if we do nothing, catastrophic change will ensue. To date, this they have utterly failed to do so.

The United Nation's Intergovernmental Panel on Climate Change is the organization that tries to convince policy makers that global warming is an immediate threat to the world. At the end of this chapter we see an image that the IPCC uses to promote the idea that its report represents the consensus view among scientists that humans are causing accelerated climate change. From the image, we can see that they are trying to make the point that 3,750 (2,500 + 800 + 450) scientists all contributed to this report and that they share the same conclusion on climate change. As we will see, this is clearly not the case.

Lawrence Solomon[13] reported an expert breakdown by Australian analyst John McLean of the contributors to the IPCC report. I will summarize it below.

First, he noticed that the great majority of the 3,750 scientists were merely reviewers. Further, the reviewers each only reviewed a small portion of the report. And he found that far from endorsing the IPCC's conclusions, many of the reviewers turned thumbs down on the IPCC sections that they read and only a handful actually endorsed the IPCC's claims that man-made global warming represents a threat to the planet.

Reviewers for the report are there to give suggestions. They can suggest changes such as rewording a poorly written passage, or they may find the data to be inconclusive or find that drawn conclusions are unfound.

In approximately 25% of the reviewer's suggestions, the editors rejected the reviewer's opinion, which allows us to see that there is far from a consensus on the IPCC report. Moreover, he found that the great majority of the reviewers commented only on chapters that dealt with historical or technical issues. These chapters did not support the IPCC's conclusions on man-made climate change.

The chapter that does deal with anthropogenic climate change is Chapter 9, "Understanding and Attributing Climate Change." Chapter 9 had 53 authors and it received comments from 55 individual reviewers. Of the 55 individual reviewers, four reviewers endorsed the entire chapter, and three reviewers endorsed a portion of the chapter.

The reviewers who endorsed the chapter were not necessarily experts in the field. For example, reviewer David Sexton stated, "Section # 9.6 I think reads pretty well for the bits I understand." This certainly does not sound like a knowledgeable person.

The 53 authors of Chapter 9 and the seven agreeable reviewers represent a total of 60 people, leading McLean to conclude, "There is only evidence that about 60 people explicitly supported the claim made by the IPCC that global warming represents a threat to the planet." 60 people hardly represent a consensus opinion.

Keep in mind that science is not a democratic process. Scientists do not vote to see which theories are accepted or rejected. Scientific theories stand on their own merits, and the existence or non-existence of a consensus in no way makes any hypothesis or theory more or less valid. Science isn't a popularity contest. Getting funding is, however.

This should give you plenty to think about the next time you hear an alarmist state that the "consensus among scientists is that humans are causing global warming." In reality, this represents a minority of scientists.

Figure 2 - This is how the IPCC report portrays its most recent report, as if there is some massive consensus among scientists that global warming is a real threat to the earth. What a joke it has become.

Chapter 3

Previous Rates Of Warming

"Without trying to prejudice this work, the results of this study will show that we know fuck-all about what the greater than 100 year variability was like with any certainty (i.e. we know with certainty that we know fuck-all)."

– Edward Cook, admitting the great uncertainty of dendrochronology.

In order to be alarming, it must appear that our most recent warming period is unique. To accomplish this goal, alarmists frequently claim that our most recent warming period occurred at a rate that has never before been matched. We will show that not only has the most recent rate of warming been matched, but that it was matched during the 20th century.

The peer-reviewed literature basically agrees that natural forces (i.e., the sun, volcanoes, etc) are strong enough to drive all of the warming and cooling that has occurred in the 4.5 billion years of the earth's existence. Alarmist scientists claim that the warming from 1975 -1998 occurred at an unparalleled rate. They use computer climate models to estimate that anthropogenic greenhouse gases are the only force strong enough to drive the warming from 1975 - 1998, and that nature and the sun played a very small role during this time period.

If we are being told that the sun was not strong enough to drive the most recent period of warming, and that human produced forces, namely carbon dioxide, were needed to achieve that rate, then that would lead you to believe that the rate of warming for that period would be greater than previously observed rates of warming. Lets see if that is true or not.

In Chapter 6 we will further discuss the below graph which was used by the Intergovernmental Panel on Climate Change (IPCC.) It will be suffice now to simply state that the graph was featured in the IPCC's 2007 Climate Assessment Report and also presented by Dr. Rajendra Pachauri (chairman of the IPCC) at the Copenhagen Conference in December 2009.

The graph below is supposed to show increasing global mean temperatures since 1860. Four separate trend lines overlie the graph, and the trend lines are supposed to represent the relative rate of warming (how quickly the earth was warming.) Each trend line is carefully crafted to give the impression that the rate of warming over the past 150 years has been accelerating.

– arguably the most global warming biased data set, more on that later) variance-adjusted global mean temperature since 1860. There are three obvious warming trends in this graph. I had the computer calculate trend lines for generally agreed upon periods of warming. These periods include 1860 - 1880, 1910 - 1940, and 1975 - 1998.

I beg the reader to understand that by plotting trend lines to show the rates of warming for the three different warming periods that I am merely employing the same logic used by the IPCC above.

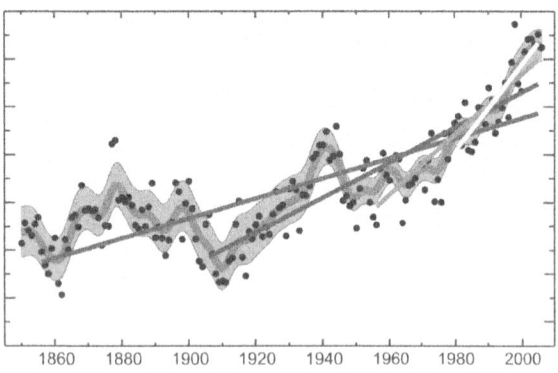

Figure 3 - Graph presented by Dr. Pachauri at the 2009 Copenhagen conference showing supposed accelerating rates of warming for recent time periods.

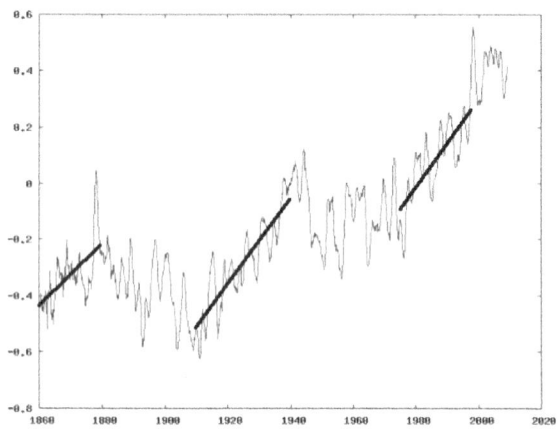

Figure 4 – Trend lines fitted to the three warming periods since 1860.

Below is a graphic showing the HadCRUT3v (a data set that is compiled by the Hadley Centre and the Climatic Research Unit from East Anglia University

Please notice that there are indeed three very distinct warming periods on this graphic. They are from 1860 - 1880, 1910 - 1940, and 1975 - 1998 (1998 is the huge peak

Previous Rates of Warming

in the graph.) Each period of warming is separated by a cooling period of approximately the same length.

Next, I calculated the slopes for each of the three trend lines:

Year Range	Slope ('C/year)	Slope ('C/decade)	Slope ('C/century)
1860 - 1880	0.0108696	0.108696	1.08696
1910 - 1940	0.0152327	0.152327	1.52327
1975 - 998	0.0154963	0.154963	1.54963

Table 1 - The calculated slopes of each trend line, represented in degrees Celsius warming per year, per decade, and per century.

It is important to note that the slope of the trend line correlates with the "rate of warming." These rates are exaggerated from what the warming rate would be if we looked at the general trend since 1850. This is because we are only looking at the slopes during warming periods and we are not looking at the vast (30 - 40 year) cooling periods in between.

Let's look at the slopes and compare them and see if the most recent warming period was at a rate greater than that of the other two periods of warming since 1860.

We can see the warming period from 1860 - 1880 has a slope of 0.0108696 or a rate of 1.08696 'C per century, which is the least steep. This indicates that this was the lowest rate of warming of the three warming periods listed above. Keep in mind that this just as the earth is coming out of a minimum of the Little Ice Age. This rate is 0.0043631 or 0.43631 'C per century less than that of 1910 - 1940, and 0.0046267 or 0.46267 'C per century less than that of 1975 - 1998.

Next, looking at 1910 - 1940, we see a slope of 0.0152327 or 1.52327 'C per century. This rate is greater than that of 1860-1880. This rate is ever so slightly less (0.0002636 or 0.02636 'C per century) than that of 1975 - 1998. This would lead one to believe that the rate of warming from 1910 - 1940 was greater than the rate of warming from 1860 - 1880, and approximately equal to the rate of warming 1975 - 1998.

Looking at 1975 - 1998, we see a slope of 0.0154963 or 1.54963 'C per century. As we saw above, the 1975 - 1998 slope is only 0.0002636 steeper than 1910 - 1940. That translates into 0.02636 'C per century, meaning that it would take 400 years to equal a difference of ~0.1'C, and 4,000 years to be ~1.0'C. The slope of 1975 - 1998 is equivocal to that of 1910 - 1940.

So what does this all mean?

Well, if we agree that there have basically been three warming periods since 1860, then we can say that two of the warming periods, 1910 - 1940 and 1975 - 1980 had approximately equivalent rates of warming, and the warming period from 1880 - 1860 occurred at a rate slightly lower than that of the other two.

Keep in mind that this slope analysis was performed on data (as you will see in Chapter 10) that has been molested as much as possible to show warming trends by raising more recent temperatures and lowering temperatures further in the past. On unmolested data, it is likely that the slope of the 1910 – 1940 warming period is greater than that of our most recent warming period.

It is interesting to note that the scientific literature agrees that natural occurrences can explain the 1910 - 1940 warming rate, when carbon dioxide levels will still pretty low. However, alarmist scientists claim that the warming from 1975 - 1998 cannot be explained by natural occurrences, despite the fact that the rate of warming was the same. Sounds fishy to me.

Chapter 4

The Medieval Warm Period

"I do not believe that global mean annual temperatures have simply cooled progressively over thousands of years as Michael Mann appears to."

– Keith Briffa, Climatic Research Unit Scientist

The Medieval Warm Period was a period of very warm climate that existed between the years AD 800 and AD 1300. During this period, the global mean temperatures were approximately 1.0′ Celsius warmer than they are currently. This period of time is the bane of alarmists. The fact that this period of time was warmer than our current period of time means that our most recent warming was not unique. If it is not unique, then there is nothing to be concerned about. The MWP was accepted as an undisputed fact for climate researchers until the idea of global warming came along and the alarmists saw that previous times were warmer than now.

In 1995 David Deming published a study in *Science* in which he looked at boreholes in North America and used them to determine that we had experienced approximately one degree Celsius of warming in the last 150 years. Alarmist scientists took this publication to mean that Deming must be a global warming believer. Shortly thereafter he testified to the U.S. Senate Committee on Environmental and Public Works that he received an email from a prominent scientist in the field that stated, "We have to get rid of the Medieval Warm Period." Deming's testimony:

"With the publication of the article in Science, I gained significant credibility in the community of scientists working on climate change. They thought I would be one of them, someone who would pervert science in the service of social and political causes. So one of them dropped his guard. An important person

working in the field of climate change and global warming sent me an astonishing email with the words: 'We must get rid of the Medieval Warm Period'"[14]

In fact it has been the tireless goal of alarmist scientists to purge this period of time. They wish to rewrite the history of the world in order to push an ideology. Millions of dollars of taxpayer money has been spent on generating graphics that minimize the Medieval Warm Period in hopes of making our current warming look unique. The most bastardly offender of this has been Penn State climatologist Michael Mann. Mann has been trying since the late 1990's to rewrite climate history and convince people that the global mean temperatures of the earth were slowly declining until around 1900 at which point he thinks that temperatures shot up rapidly.

THE HOCKEY STICK

Below we can see a graphic containing the first major graphic that Mann published in attempt to belittle the MWP. Notice that the graphic trends downward, slowly, for nearly a thousand years, until around 1900 when the graphic rapidly spikes upwards. This is called a "hockey-stick" graph, because the whole graph is straight until the end, so it looks similar to a hockey stick laid out on the ground. This graphic is meant to give the impression that the world's temperatures have declined slowly over thousands of years, and then, suddenly, they rose rapidly.

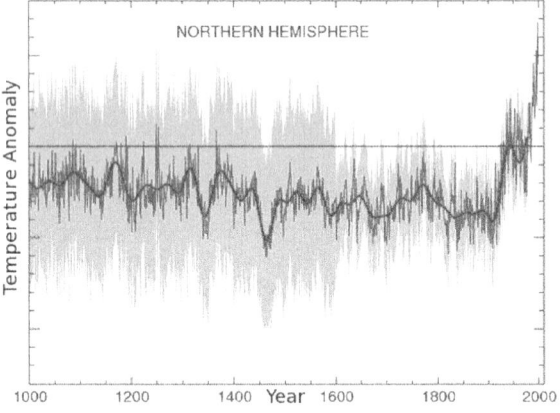

Figure 5 - Michael Mann's original (now debunked) hockey stick graph.

Figure 6 - A real hockey stick. Compare the shape of this to Mann's graph to understand its namesake.

Compare Mann's hockey-stick graph (Figure 5) to the version of the last 1,000 years that actually represents reality (Figure 7.)

The Medieval Warm Period

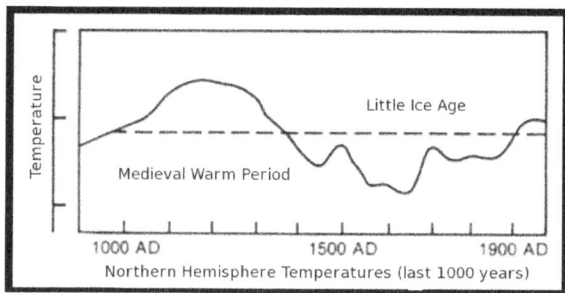

Figure 7 - 1,000 years of temperatures, including the Medieval Warm Period and the Little Ice Age. This graph was originally presented by the IPCC before the "hockey team" came along.

Notice that Mann's graphic does not show any large spike in temperatures during the Medieval Warm Period – it has simply vanished. And notice that there is no dramatic drop in temperatures during the Little Ice Age. It has simply vanished as well!

Now, it is important to understand how Mann got this data for the last 1,000 years. We will briefly discuss his methodologies, but first we must have a basic understanding of dendrochronology.

Dendrochronology (dendro = tree; chron = time; ology = study of) is the scientific method of dating based on the analysis of patterns of tree-rings. Dendrochronologists study tree rings and attempt to reconstruct past temperature and climate data from them. Since we did not have accurate thermometer records prior to 1850-ish we have to rely on other sources of data for temperature information. These other sources are called proxies.

Proxy records contain a signal that corresponds to climate, but that signal may be weak and embedded in a great deal of background noise. Deciphering that record is often complicated. Examples of proxy records include ice cores, lake sediments, ocean sediments, and tree rings.

To determine the past temperature, the dendrochronologist cuts a core sample out of a tree, extracts it, and then looks at the width of the tree rings. They trust that a wider tree ring indicates a greater growing season, which they believe is primarily driven by a higher temperature. This method has many faults that are further discussed in detail in Chapter 11.

It is important to note that Michael Mann is not a dendrochronologist. He uses other people's data and applies statistics (poorly, as we will see) to them in order to get the results he is looking for.

In 1998 Michael Mann, Raymond Bradley, and Malcolm Hughes (MBH98) published the original hockey stick graph that we previously saw. At the time, this graph received a lot of media attention because it really supported the idea of catastrophic anthropogenic global warming. Remember that the graph showed relatively stable temperatures for a thousand years, and then a sharp incline in recent times, indicating that humans have caused this sharp change in global temperatures. The high-profile publication of the data led to the "hockey stick" being used as a key piece of supporting evidence in the Third Assessment Report by the United Nations' Intergovernmental Panel on Climate Change (IPCC) in 2001.

An email from Michael Mann demonstrates his desire to get rid of this "damning" period of very warm temperatures:

"Raymond Bradley, Keith Briffa, Tom Crowley, Phil Jones, Michael Oppenheimer, Jonathan Overpeck, Kevin Trenberth, Tom Wigley,

I think that trying to adopt a timeframe of 2000 years, rather than the usual 1000 years, addresses a good earlier point that Jonathan Overpeck made ... that it would be nice to try to "contain" the putative "Medieval Warm Period", even if we don't yet have data available that far back."

- Michael Mann[15]

DEBUNKING THE HOCKEY STICK

In 2003, Stephen McIntyre and Ross McKitrick published a paper titled "Corrections to the Mann et al. (1998) Proxy Data Base and Northern Hemisphere Average Temperature Series" in the journal Energy and Environment 14(6) 751-772, raising concerns about their inability to reproduce the results that Mann and his co-authors obtained in their original hockey stick graph.

The abstract for their paper reads:

"The data set of proxies of past climate used in Mann, Bradley and Hughes (1998, "MBH98" hereafter) for the estimation of temperatures from 1400 to 1980 contains collation errors, unjustifiable truncation or extrapolation of source data, obsolete data, geographical location errors, incorrect calculation of principal components and other quality control defects. We detail these errors and defects. We then apply MBH98 methodology to the construction of a Northern Hemisphere average temperature index for the 1400-1980 period, using corrected and updated source data. The major finding is that the values in the early 15th century exceed any values in the 20th century. The particular "hockey stick" shape derived in the MBH98 proxy construction – a temperature index that decreases slightly between the early 15th century and early 20th

century and then increases dramatically up to 1980 – is primarily an artifact of poor data handling, obsolete data and incorrect calculation of principal components."

Basically, they found errors in Mann's methods and fixed them. Specifically, McKitrick comments:

"The Mann mult-iproxy data, when correctly handled, shows the 20th century climate to be unexceptional compared to earlier centuries."

In the McIntyre and McKitrick paper they applied corrected statistics to Mann's data and produced the following graph:

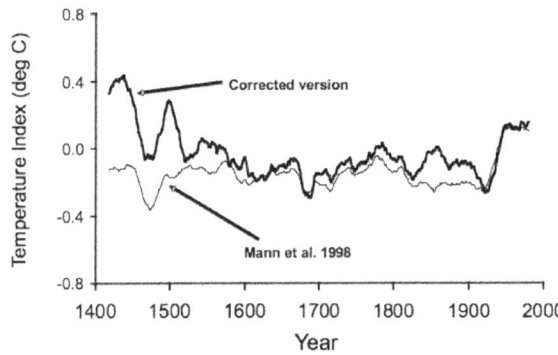

Figure 8 - McIntyre and McKitrick corrections (showing warmer past temperatures) to Mann's original data (showing lower temperatures.)

Clearly, the corrected version is not as alarming as Mann's original. The corrected data plainly demonstrates that temperatures in the past were higher than they are currently. This tells us that our current warming period is not unique. And we certainly were not producing excessive amounts of greenhouse gases one thousand years ago to obtain those temperatures.

In fact, it was found that random red noise (different than white noise) could be fed into Michael Mann's methodologies and it would produce a hockey stick shaped graph. Basically, any data that was fed into Mann's code would make the same graph shape! His results were meaningless. Your tax dollars paid for this shoddy work.

This debunking of Mann's hockey stick graph caused a huge uproar in the climatology community. The ClimateGate scientists were not used to people questioning their data or methods.

In fact, Michael Mann received a secret email forward before McIntyre & McKitrick's paper was published letting him know of the impending publication:

"Michael Mann,

Two people have a forthcoming 'Energy & Environment' paper that's being unveiled tomorrow that will claim that Mann arbitrarily ignored paleoclimate data within his own record and substituted other data for missing values that dramatically affected his results.

When his exact analysis is rerun with all the data and with no data substitutions, two very large warming spikes will appear that are greater than the 20th century.

Personally, I'd offer that this was known by most people who understand Mann's methodology. It can be quite sensitive to the input data in the early centuries.

Anyway, there's going to be a lot of noise on this one, and knowing Mann's very thin skin I am afraid he will react strongly, unless he has learned (as I hope he has) from the past...."

– *Unknown Sender*[16]

I find it rather disturbing that "this was known by most people who understand Mann's methodology." If it is known in the climate science community that Mann arbitrarily ignores data within his own record and substitutes other data for missing values that dramatically affects his results, how do his papers pass peer-review?

Not surprisingly, Michael Mann forwards his secret email to seventeen of his colleagues along with the following:

"To All,

This [the secret email from above] has been passed along to me by someone whose identity will remain in confidence. Its clear that "Energy and Environment" is being run by the baddies.

My suggested response is:

1) to dismiss this as stunt, appearing in a so-called "journal" which is already known to have defied standard practices of peer-review. It is clear, for example, that nobody we know has been asked to "review" this so-called paper

2) to point out the claim is nonsense since the same basic result has been obtained by numerous other researchers, using different data, elementary compositing techniques, etc.

Who knows what sleight of hand the authors of this thing have pulled. Of course, the usual suspects are going to try to peddle this crap. The important thing is to deny that this has any intellectual credibility whatsoever and, if contacted by any media, to dismiss this for the stunt that it is."

- *Michael Mann*[17]

Starting from the beginning of his email, we first see that Mann will attack and make accusations about any scientific journal that publishes papers that are not in line with his ideological views. The McIntyre & McKitrick paper's results were not in line with what Mann believes, so he insults the journal and accuses it of being run by "the baddies."

Next, we see that he says his colleagues should dismiss the publication of the dissenting paper as a stunt. Mann displays incredible arrogance in assuming that every paper submitted for publication by a journal should automatically be passed to one of his colleagues. If this had previously been the case, can you imagine how a select few could control what is published by science and what isn't?

After that, we see that Mann tells his colleagues to point out that the claim in the McIntyre & McKitrick paper is nonsense and has no intellectual credibility. But Mann has not even read the paper yet! Keep in mind that he had only just received a heads up telling him that a paper was being published that points out a number of errors in MBH98.

One would think that a scientist would want to read a paper before telling their colleagues that it is all nonsense – especially when his colleagues are aware that Mann uses fallible methodologies! Mann sounds more like a guilty child than a scientist.

Shortly after, Raymond Bradley fired off an email to his friends at the Climatic Research Unit of East Anglia University, begging them to write something up to squash McIntyre & McKitrick's debunking of Mann's hockey stick.

"Phil Jones, Keith Briffa, Tim Osborn,

I suggest a way out of this mess. Because of the complexity of the arguments involved, to an uniformed observer it all might be viewed as just scientific nit-picking by "for" and "against" global warming proponents. However, if an "independent group" such as you guys at CRU could make a statement I think that would go a long way to defusing the issue.

If you are willing, a quick and forceful statement from The Distinguished CRU Boys would help quash further arguments, although here, at least, it is already quite out of control.

Yesterday in the U.S. Senate the debate opened on the McCain-Lieberman bill to control CO2 emissions from power plants. Senator Inhofe stood up & showed the McIntyre & McKitrick figure and stated that MBH98 & the IPCC assessment was now disproved and so there was no reason to control CO2 emissions."

- Ray Bradley[18]

Later, Tom Wigley writes this incredibly damning letter to Phil Jones. Wigley agrees with the skeptics that Mann's paper is rubbish.

"Phil Jones,

I have just read the McIntyre & McKitrick stuff criticizing Mann's paper. A lot of it seems valid to me. At the very least MBH is a very sloppy piece of work -- an opinion I have held for some time.

Presumably what you have done with Keith is better -- or is it? I get asked about this a lot. Can you give me a brief heads up?

Mike is too deep into this to be helpful."

- Tom Wigley

The controversy over the debunking of Mann's hockey stick graph continued. In 2006 a team of statisticians led by Edward Wegman, chair of the National Academy of Sciences' (NAS) Committee on Applied and Theoretical Statistics, was assembled at the request of U.S. Representative Joe Barton and U.S. Representative Ed Whitfield. They were asked to study and then write a report which primarily focused on the statistical analysis used in the MBH98 paper, and also considered the personal and professional relationships between Mann et al. and other members of the paleoclimate community. Findings presented in this report (commonly known as the "Wegman Report") at a hearing of the subcommittee on oversight and investigations, chaired by Whitfield, included the following:

"Mann et al., misused certain statistical methods in their studies, which inappropriately produce hockey stick shapes in the temperature history. Wegman's analysis concludes that Mann's work cannot support claim that the 1990s were the warmest decade of the millennium.

Report: "Our committee believes that the assessments that the decade of the 1990s was the hottest decade in a millennium and that 1998 was the hottest year in a millennium cannot be supported by the MBH98/99 analysis. As mentioned earlier in our background section, tree ring proxies are typically calibrated to remove low frequency variations. The cycle of Medieval Warm Period and Little Ice Age that was widely recognized in 1990 has disappeared from the MBH98/99 analyses, thus making possible the hottest decade/hottest year claim. However, the methodology of MBH98/99 suppresses this low frequency information. The paucity of data in the more remote past makes the hottest-in-a-millennium claims essentially unverifiable."

A social network analysis revealed that the small community of paleoclimate researchers appears to review each other's work, and reuse many of the same data sets, which calls into question the independence of peer- review and temperature reconstructions.

Report: "It is clear that many of the proxies are re-used in most of the papers. It is not surprising that the papers would obtain similar results and so cannot really claim to be independent verifications."

Although the researchers rely heavily on statistical methods, they do not seem to be interacting with the statistical community.

Report: "As statisticians, we were struck by the isolation of communities such as the paleoclimate community that rely heavily on statistical methods, yet do not seem to be interacting with the mainstream statistical community. The public policy implications of this debate are financially staggering and yet apparently no independent statistical expertise was sought or used."

Authors of policy-related science assessments should not assess their own work.

Report: "Especially when massive amounts of public monies and human lives are at stake, academic work should have a more intense level of scrutiny and review. It is especially the case that authors of policy-related documents like the IPCC report, Climate Change 2001: The Scientific Basis, should not be the same people as those that constructed the academic papers."

Policy-related climate science should have a more intense level of scrutiny and review involving statisticians. Federal research should involve interdisciplinary teams to avoid narrowly focused discipline research.

Report: "With clinical trials for drugs and devices to be approved for human use by the FDA, review and consultation with statisticians is expected. Indeed, it is standard practice to include statisticians in the application-for-approval process. We judge this to be a good policy when public health and also when substantial amounts of monies are involved, for example, when there are major policy decisions to be made based on statistical assessments. In such cases, evaluation by statisticians should be standard practice. This evaluation phase should be a mandatory part of all grant applications and funded accordingly."

Federal research should emphasize fundamental understanding of the mechanisms of climate change, and should focus on interdisciplinary teams to avoid narrowly focused discipline research.

Report: "While the paleoclimate reconstruction has gathered much publicity because it reinforces a policy agenda, it does not provide insight and understanding of the physical mechanisms of climate change... What is needed is deeper understanding of the physical mechanisms of climate change."

Further, the report claimed that the MBH98 method creates a hockey-stick shape even when supplied with random input data, and argues that the MBH98 method uses weather station data from 1902 to 1995 as a basis for calibrating other input data. "It is not clear that Dr. Mann and his associates even realized that their methodology was faulty at the time of writing the MBH98 paper. The net effect of the de-centering [a statistical method Mann employed] is to preferentially choose the so-called hockey stick shapes."

Now, I want to be fair. Not all climate scientists believed Mann's idea that the earth cooled progressively over thousands of years and then warmed suddenly. In fact Mann's idea is rather dumb and it is actually surprising that his paper would get past peer-review. Privately, other climate scientists expressed their disagreement with Mann's ideas:

"Let me say that I don't mind what you put in the policy makers summary if there is a general consensus. However some general discussion would be valuable ... whether this represents 'TRUTH' however is a difficult problem. I know Mike thinks his series is the 'best' and he might be right - but he may also be too dismissive of other data and possibly over confident in his (or should I say his use of other's...

I know there is pressure to present a nice tidy story as regards 'apparent unprecedented warming in a thousand years or more in the proxy data' but in reality the situation is not quite so simple. We don't have a lot of proxies that come right up to date and those that do (at least a significant number of tree proxies) some unexpected changes in response that do not match the recent warming. I do not think it wise that this issue be ignored in the chapter...

I believe that the recent warmth was probably matched about 1000 years ago. I do not believe that global mean annual temperatures have simply cooled progressively over thousands of years as Mike appears to and I contend that that there is strong evidence for major changes in climate over the Holocene that require explanation and that could represent part of the current or future background variability of our climate..."

– Keith Briffa[19]

In 2009 Mann et al. published a new paper[20] in which they found that the Medieval Warm Period shows:

"Warmth that matches or exceeds that of the past decade in some regions, but which falls well below recent levels globally."

It must have killed Mann to write that. He contends that the Medieval Warm Period occurred only regionally, and was not a global phenomenon. Their new reconstruction of that period was characterized by warmth over large parts of the North Atlantic, Southern Greenland, the Eurasian Arctic, and parts of North America which appeared to substantially exceed that of modern late 20th century (1961–1990) baseline and was comparable to or exceeded that of the past one-to-two decades in some regions.

It took Michael Mann more than 10 years and millions of taxpayer dollars to arrive at the same conclusion that we already have had for many decades. What a monumental waste of money and time.

OTHER SUPPORTING DATA

Fortunately, there are many other scientists that are interested in reconstructing temperatures that encompass the last 1,000 years. In these reconstructions, the vast majority demonstrates that temperatures during the Medieval Warm Period were warmer than they are currently.

Soon[21] and Baliunas performed a review of the paleoclimate literature and found more than 200 studies which concluded, "The 20th century is probably not the warmest nor a uniquely extreme climatic period of the last millennium." Clearly Mann's idea of slowly decreasing temperatures for a thousand years is not the accepted view. But it sure is the most political.

The Center for the Study of Carbon Dioxide and Global Change[22] has done exhaustive reviews of peer-reviewed papers that discuss the MWP. Their findings state that according to data published by 794 individual scientists from 473 separate research institutions in 42 different countries the Medieval Warm Period existed and was warmer than current temperatures. A qualitative distribution of the studies they reviewed is shown below:

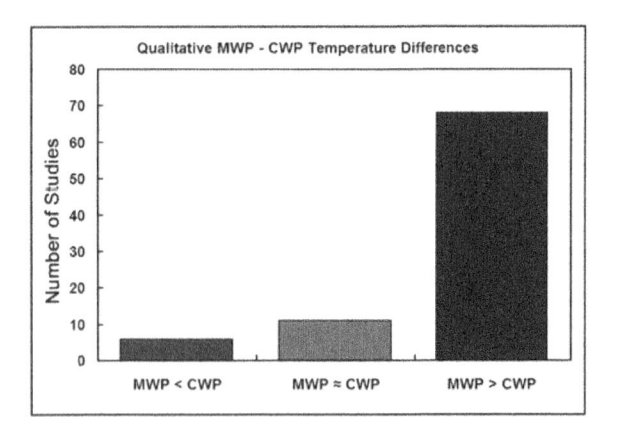

Figure 9 - Courtesy of co2science.org. MWP = Medieval Warm Period; CWP = Current Warm Period. This graph shows the majority of studies find that the MWP was warmer than the CWP.

In the above graphic, the first bar represents studies that show the Medieval Warm Period was cooler than the Current Warm Period (CWP – our current temperatures.) As you can see this is the smallest representation and is the minority of studies. Mann's study would be included in this part.

The next bar represents studies that conclude that the Medieval Warm Period was equivalent in temperature to the Current Warm Period.

The last bar represents studies that state that the Medieval Warm Period was warmer than the Current Warm Period. Clearly, these studies represent the majority.

Briefly, we can look at a chart that demonstrates quantitatively how much warmer the MWP was than the CWP according to the same studies:

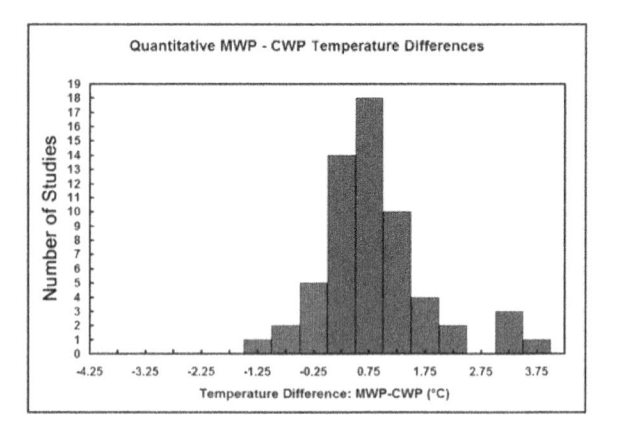

Figure 10 - Courtesy of co2science.org. A quantitative comparison of MWP and CWP temperature differences. This graph shows how much warmer studies find the MWP being than the CWP.

We will now look at a sample of three of the studies that are included in the above graphics.

The first study comes from Tyson et al.[23] Courtesy of CO2Science, "In this study the maximum annual air temperatures in the vicinity of Cold Air Cave (24°1'S, 29°11'E) in the Makapansgat Valley of South Africa were inferred from a relationship between color variations in banded growth-layer laminations of a well-dated stalagmite and the air temperature of a surrounding 49-station climatological network developed over the period 1981 - 1995, as well as from a quasi-decadal-resolution record of oxygen and carbon stable isotopes. In this study the Medieval Warm Period (AD 1000 - 1325) was found to be as much as 3 - 4°C warmer than the Current Warm Period (AD 1961 - 1990 mean)."

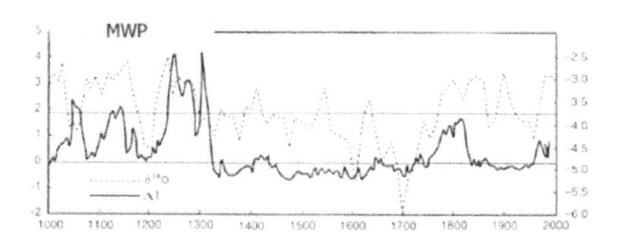

Figure 11 - Courtesy of co2science.org. Makapansgat Valley proxy temperature reconstruction adapted from Tyson et al.

The next study comes from Goni et al.[24] Courtesy of CO2Science, "Based on the degree of unsaturation of certain long-chain alkenones synthesized by haptophyte algae contained in a sediment core retrieved from the eastern sub-basin of the Cariaco Basin (10°30'N, 64°40'W) on the continental shelf off the Venezuelan central coast, Goni et al. determined that the highest sea surface temperatures at that location over the past 6000 years 'were measured during the Medieval Warm Period (MWP),' which they identified as occurring between AD 800 and 1400. From the graph of their results reconstructed below, it is further evident that peak MWP temperatures were approximately 0.35°C warmer than peak Current Warm Period temperatures, and that they were fully 0.95°C warmer than the mean temperature of the last few years of the 20th century."

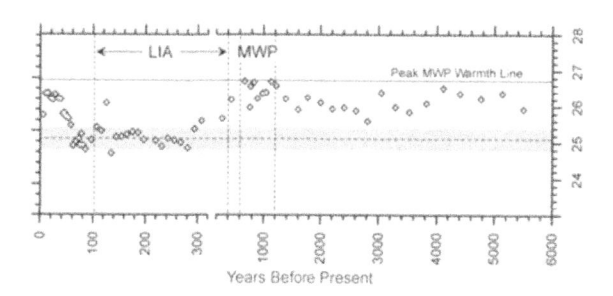

Figure 12 - Courtesy of co2science.org. Alkenone-based Sea Surface Temperature reconstruction for the Cariaco Basin, Venezuela, showing temperatures of the Medieval Warm Period were higher than those of the Current Warm Period. Note that the X-axis (years) is reversed from what you may expect. It shows years before present, so the most recent times are on the left, with the times further in the past on the right.

The last study comes from von Gunten et al.[25] Courtesy of CO2Science, "Von Gunten et al. developed a continuous high-resolution (1-3 years sampling interval, 5-year filtered reconstruction) austral summer (December to February) temperature reconstruction based on chloropigments derived from algae and phototrophic bacteria found in sediment cores retrieved from Central Chile's Laguna Aculeo (33°50'S, 70°54'W) in 2005 that extended back in time to AD 850. This work provided, in their words, "quantitative evidence for the presence of a Medieval Warm Period (a.k.a. Medieval Climate Anomaly) (in this case, warm summers between AD 1150 and 1350; ΔT = +0.27 to +0.37°C with respect to twentieth century) and a very cool period synchronous to the 'Little Ice Age' starting with a sharp drop between AD 1350 and AD 1400 (-0.3°C/10 years, decadal trend) followed by constantly cool (ΔT = -0.70 to -0.90°C with respect to twentieth century) summers until AD 1750."

Looking at the graph of their data, the peak warmth of the Current Warm Period (CWP) occurred in the late 1940s. Since that time, temperatures have declined and then risen, but not to the level of warmth experienced earlier in the century. Peak warmth of the MWP is about 0.5°C higher than that recorded for the past two decades of the 20th century, which is claimed by the world's climate alarmists to have been the warmest of the past thousand or more years. Hence, it is this latter period to which we compare the peak warmth of the MWP.

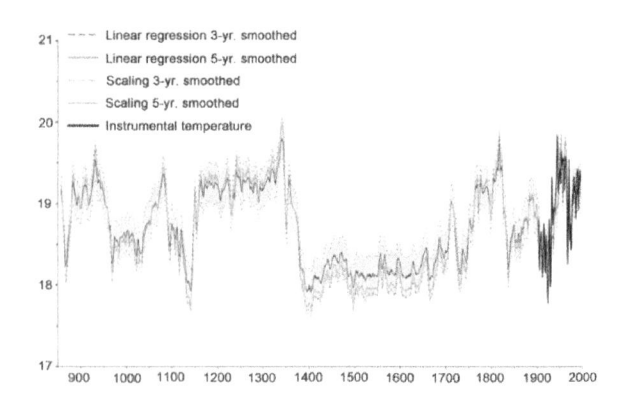

Figure 13 - Courtesy of co2science.org. Von Gunten et al. graph showing the warmth of the MWP being greater than the warthm of the CWP.

In addition to this overwhelming body of evidence, we can briefly look at some archeological evidence that supports the existence of the Medieval Warm Period.

The Vikings[26] were Norse (Scandinavian) explorers, warriors, merchants, and pirates who raided and colonized wide areas of Europe from the late eighth to the early eleventh century. The Vikings took advantage of ice-free seas during the Medieval Warm Period to colonize Greenland and other outlying lands of the far north. Around 1000 AD the climate was sufficiently warm for the north of Newfoundland to support a Viking colony.

The Viking settlements flourished for hundreds of years until the world fell into the Little Ice Age and the dramatically cooling climate made life in Greenland inhospitable. Without the Medieval Warm Period the Vikings would not have been able to easily navigate the ice-free seas, and they would not have been able to settle Greenland and have it flourish.

The Vikings were originally pagans. However, Olav Tryggvason came from England in the summer of AD 995 to claim the throne of Norway and to bring Christianity to the country. On the island of Moster Tryggvason held the first official Christian mass in Norway. Approximately 35 years later Norway was officially a Christian country.

This is fortunate for archeologists because the Vikings' conversion to Christianity meant that they now buried their dead instead of burning them. During archeological studies in Greenland Viking corpses have been found buried under what is now permafrost. This means that the ground was warmer during the past (enough so to melt the permafrost) allowing the corpses to be buried, and our current temperatures are not warm enough to melt the permafrost.

Further, we have extensive evidence of global glacial retreat during the period between AD 900 and 1300.[27] This is to be expected during warmer periods. Ice forms during colder periods and melts during warmer periods. Interestingly, glaciers that have retreated since 1850 uncover plant leftovers from the Middle Ages, which is clear evidence that the extent of the glaciers at that time was lower than today.[28] [29]

Lastly, the growth of crops can also demonstrate the warmth of the Medieval Warm Period. The Alps tree line climbed 2000 meters higher than what we see from the current tree levels.[30] Winery was possible in Germany at the Rhine and Mosel up to 200 meters above the present limits, in Pomerania, East Prussia, England and southern Scotland, and in southern Norway, therefore, much farther north than is possible today.[31] In many parts of the world farmable land reached heights that were never reached again later.[32]

Despite the best efforts of the ClimateGate scientists, they have been unable to hide the Medieval Warm Period. It is supported by an overwhelming abundance of evidence across multiple scientific disciplines. The Medieval Warm Period unequivocally demonstrates that warm climates are not unique and that it was warmer in the past without the presence of human produced greenhouse gases.

Chapter 5

The Solar Connection

"In 20 years, the West Side Highway [on the western side of Manhattan Island] will be under water. And there will be tape across the windows across the street because of high winds. And the same birds won't be there. The trees in the median strip will change."

- James Hansen, Father of Global Warming, in 1988

The Sun[33] is the huge flaming ball of gas at the center of our solar system. It has a diameter of about 1,392,000 kilometers (about 109 Earths), and by itself accounts for about 99.86% of our solar system's mass.

It generates energy by nuclear fusion of hydrogen nuclei into helium. The Sun's hot corona continuously expands in space creating the solar wind, a hypersonic stream of charged particles that extends roughly 100 AU (1 AU = 1 astronomical unit = the distance from the earth to the sun = ~149,600,000 kilometers = ~93 million miles.)

The energy of the sunlight ultimately supports almost all life on Earth via photosynthesis, and drives Earth's climate and weather. The enormous impact of the Sun on the Earth has been recognized since pre-historic times, and the Sun has been regarded by some cultures as a deity.

The Sun is a magnetically active star. It supports a strong, changing magnetic field that varies year-to-year and reverses direction about every eleven years around solar maximum (periods of maximal solar output.) The Sun's magnetic field gives rise to many effects that are collectively called solar activity, including sunspots on the surface of the Sun, solar flares, and variations in solar wind that carry material through the Solar System.

Effects of solar activity on Earth include auroras at moderate to high latitudes, and the disruption of radio communications and electric power. Solar activity also has impacts on the structure of Earth's outer atmosphere.

All matter in the Sun is in the form of gas and plasma because of its high temperatures. This makes it possible for the Sun to rotate faster at its equator (about 25 days) than it does at higher latitudes (about 35 days near its poles). The differential rotation of the Sun's latitudes causes its magnetic field lines to become twisted together over time, causing magnetic field loops to erupt from the Sun's surface and trigger the formation of the Sun's dramatic sunspots and solar prominences. This twisting action gives rise to the solar magnetic field and an 11-year solar cycle of magnetic activity as the Sun's magnetic field reverses itself about every 11 years.

When observing the Sun with appropriate filtration, the most immediately visible features are usually its sunspots, which are well-defined surface areas that appear darker than their surroundings because of lower temperatures. Sunspots are regions of intense magnetic activity where convection is inhibited by strong magnetic fields, reducing energy transport from the hot interior to the surface. The magnetic field gives rise to strong heating in the corona, forming active regions that are the source of intense solar flares and coronal mass ejections. The largest sunspots can be tens of thousands of kilometers across.

These solar flares and coronal mass ejections cause an increase in the solar output. This translates into greater solar forcing upon the climate. Just to make sure we have it straight: more sunspots = more solar forcing = warmer weather. Less sunspots = less solar forcing = cooler weather.

The number of sunspots visible on the Sun is not constant, but varies over an 11-year cycle known as the solar cycle. At a typical solar minimum, few sunspots are visible, and occasionally none at all can be seen. Those that do appear are at high solar latitudes. As the sunspot cycle progresses, the number of sunspots increases and they move closer to the equator of the Sun, a phenomenon described by Spörer's law. Sunspots usually exist as pairs with opposite magnetic polarity. The magnetic polarity of the leading sunspot alternates every solar cycle, so that it will be a north magnetic pole in one solar cycle and a south magnetic pole in the next.

The solar cycle has a great influence on space weather, and is a significant influence on the Earth's climate since luminosity has a direct relationship with magnetic activity.

The Solar Connection

Solar activity minima tend to be correlated with colder temperatures, and longer than average solar cycles tend to be correlated with hotter temperatures.

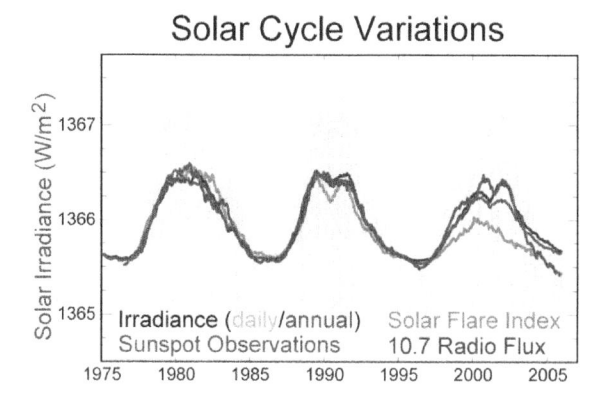

Figure 14 - Graphic demonstrating the 11-year sunspot cycle.

In the 17th century, the solar cycle appears to have stopped entirely for several decades; very few sunspots were observed during this period. This period of missing sunspots is known as the Maunder Minimum[34] and it coincides with the Little Ice Age. Earlier extended minima have been discovered through analysis of tree rings and also appear to have coincided with lower-than-average global temperatures. The connection between sunspots and temperatures is very strong.

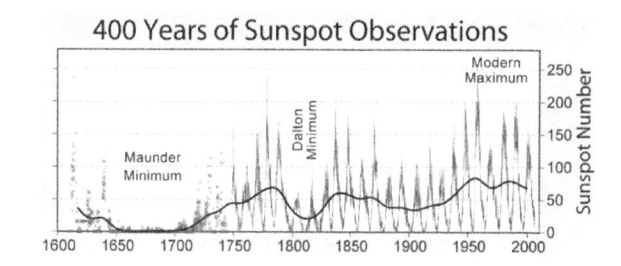

Figure 15 - Graphic demonstrating the Maunder Minimum coinciding with the same period of time as the Little Ice Age.

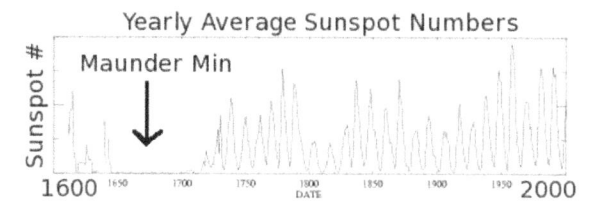

Figure 16 - Another graphic demonstrating the Maunder Minimum coinciding with the same period of time as the Little Ice Age.

Above, we can clearly see the almost complete absence of sunspots during the Maunder Minimum occurring during the same time period as the Little Ice Age. Additionally, we can see that after the Maunder Minimum the number of sunspots slowly increased over time – and with that temperatures slowly rose over time. In the most recent part of the graph we can see the modern maximum of sunspots, which just so happens to coincide with our recent warming period.

The Maunder Minimum[35] is the name used for the period roughly spanning AD 1645 to 1715 by John A. Eddy in a landmark 1976 paper published in *Science* titled "The

Maunder Minimum", when sunspots became exceedingly rare, as noted by solar observers of the time. Astronomers before Eddy had also named the period after the solar astronomer Edward W. Maunder (1851–1928) who studied how sunspot latitudes changed with time. The periods he examined included the second half of the 17th century. Edward Maunder published two papers in 1890 and 1894, and he cited earlier papers written by Gustav Spörer. The Maunder Minimum's duration was derived from Spörer's work. Like the Dalton Minimum and Spörer Minimum, the Maunder Minimum coincided with a period of lower-than-average global temperatures.

During one 30-year period within the Maunder Minimum, astronomers observed only about 50 sunspots, as opposed to a more typical 40,000–50,000 spots in modern times.

The Maunder Minimum coincided with the middle — and coldest part — of the Little Ice Age, during which Europe and North America, and perhaps much of the rest of the world, were subjected to bitterly cold winters.

In 1709, the Rhine remained frozen until the summer. This contributed to widespread starvation and the lengthy emigration of Germans from the Palatine. These people are referred to as the Poor Palatines.

During the Great Frost[36] of 1683 - 1684, the worst frost recorded in England, the Thames was completely frozen for two months. Solid ice was reported extending for miles off the coasts of the southern North Sea (England, France and the Low Countries), causing severe problems for shipping and preventing the use of many harbors.

Sunspots have been counted for thousands of years. In the modern world, sunspots are observed with land-based and Earth-orbiting solar telescopes. These telescopes use filtration and projection techniques for direct observation, in additional to various types of filtered cameras. Specialized tools such as spectroscopes and spectrohelioscopes are used to examine sunspots and sunspot areas. Artificial eclipses allow viewing of the circumference of the sun as sunspots rotate through the horizon.

The Solar Connection

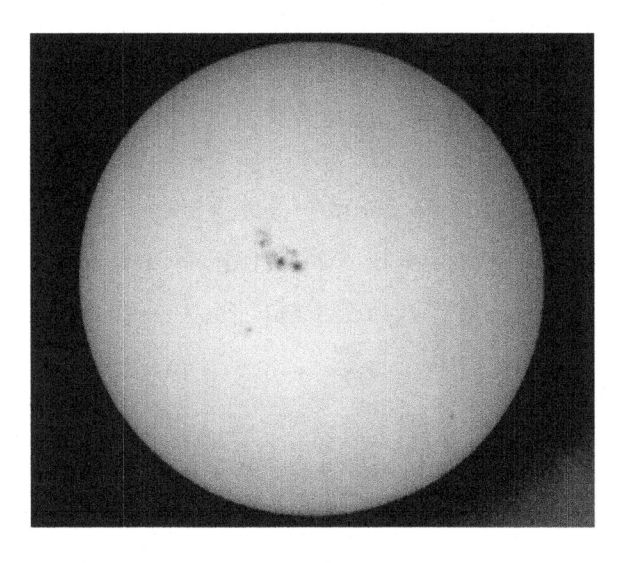

Figure 17 - An example of a sunspot on the sun.

Sunspot numbers over the past 11,400 years have been reconstructed using dendrochronologically dated radiocarbon concentrations. The level of solar activity during the past 70 years is exceptional — the last period of similar magnitude occurred over 8,000 years ago. The Sun was at a similarly high level of magnetic activity for only ~10% of the past 11,400 years, and almost all of the earlier high-activity periods were shorter than the present episode.

Is it any surprise we have had warming recently?

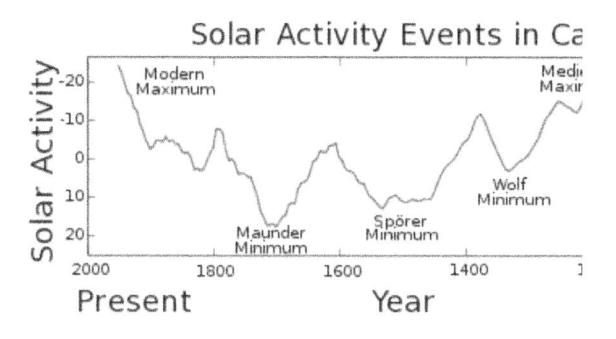

Figure 18 - Sunspot numbers over the last 2,000 years reconstructed from radiocarbon concentrations.

Solanki et al. 2003[37] found that the unusually active sun in the last 60 years is unique in the last millennium. They performed a reconstruction of sunspot numbers using the radioactive isotope of Beryllium, Beryllium-10. The reconstruction shows reliably that the period of high solar activity during the last 60 years is unique throughout the past 1150 years. This nearly triples the time interval for which such a statement could be made previously.

Their sunspot reconstruction is shown below:

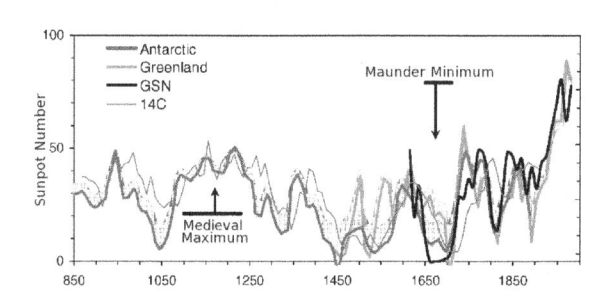

Figure 19 - Solanki et al. 2003 sunspot reconstruction using Beryllium-10 radioisotopes.

The most striking feature of their complete sunspot number profile is the uniqueness of the steep rise of sunspot activity during the first half of the 20th century. Never during the 11 centuries prior to that was the Sun nearly as active.

This may initially sound odd, because we are aware of the Medieval Maximum, a period around a thousand years ago when the sun was very active, which coincided with the Medieval Warm Period. Let's look further into this.

While the average value of the reconstructed sunspot number between 850 and 1900 is about 30, it reaches values of 60 since 1900 and 76 since 1944. For the observed group sunspot series since 1610 these values are 25, 61, and 75, respectively. The largest 100-year average of the reconstructed sunspot numbers prior to 1900 is 44, which occurs in 1140–1240, that is, during the medieval maximum, but even this is significantly less than the level reached in the last century.

The medieval maximum is remarkable, however, in the length of time that the Sun consistently remained at the average sunspot number level of about 40–50. Only during the recent period of high activity since about 1830 has the sunspot number remained consistently above 30 for a similar length of time. The consistently high sunspot number during the Medieval Maximum accounts for the higher temperatures of the Medieval Warm Period.

Solanki concluded that the high level of solar activity since the 1940s is unique since the year 850. This can be considered a robust conclusion since they have shown that their reconstruction is particularly reliable in phases of high and intermediate sunspot activity, while during periods of low activity the sunspot number may be overestimated.

Clearly, we can see that the sun is the dominant force driving climate change.

Chapter 6

Recent Cooling

"The scientific community would come down on me in no uncertain terms if I said the world had cooled from 1998. OK it has but it is only 7 years of data and it isn't statistically significant."

– Phil Jones, July 2005

One of the most detrimental facts for human caused global warming is that we have been experiencing global cooling since 2001/2002 (some say since 1998; there is also evidence that we have not had statistically significant global warming since 1995.) Alarmists would like for this recent cooling to not exist because it contradicts the idea that "runaway carbon dioxide levels" are causing unparalleled rises in temperatures. Dropping temperatures, despite rising carbon dioxide levels surely must make them scratch their head.

THE IPCC GRAPH

In this chapter I am going to demonstrate using the same procedure as the Intergovernmental Panel on Climate Change (IPCC) that there has been global cooling since 2001/2002.

On the next page is a graph that was featured in the IPCC's 2007 Climate Assessment Report. We have seen this graphic before. It shows the slow rise in temperatures over the last 150 years. Four separate trend lines overlie it, each with a start date carefully selected to give the impression that the rate of warming over the past 150 years has been accelerating. The slope of the trend line is supposed to correlate with the rate at which warming is occurring; therefore a steeper slope indicates a greater rate of warming.

This same graph was presented again by Dr. Rajendra Pachauri, chairman of the IPCC (who has a background as a railroad engineer), at the Copenhagen conference on climate in December 2009. This graph was also featured in Lord Monckton's rather damning letter[38] to Dr. Pachauri:

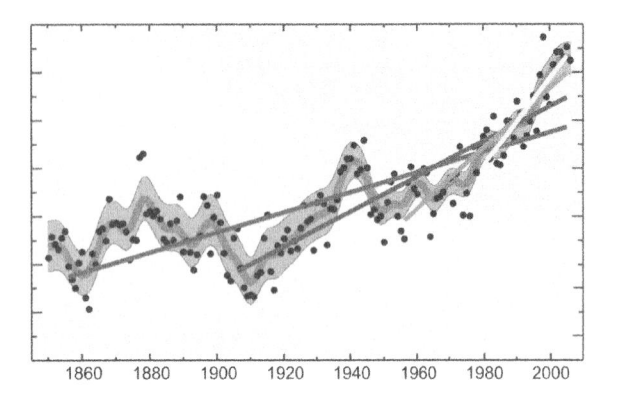

Figure 20 – Graph presented by Dr. Pachauri at the 2009 Copenhagen conference showing supposed accelerating rates of warming for recent time periods.

As you can see, as the start-date for each trend line is carefully moved closer to the present we can see an increase in the slope of the line. Again, this is supposed to correlate with an increased rate of warming in more recent times. (For fun, also note that the trend lines sort of "hide the decline" of temperatures around 2000.)

Below I've reconstructed their graph, showing the same 150, 100, 50, and 25-year trends of warming. I have done this so that the reader can see that the computer graphing and trend-line generating procedure I use generates the same results as the IPCC:

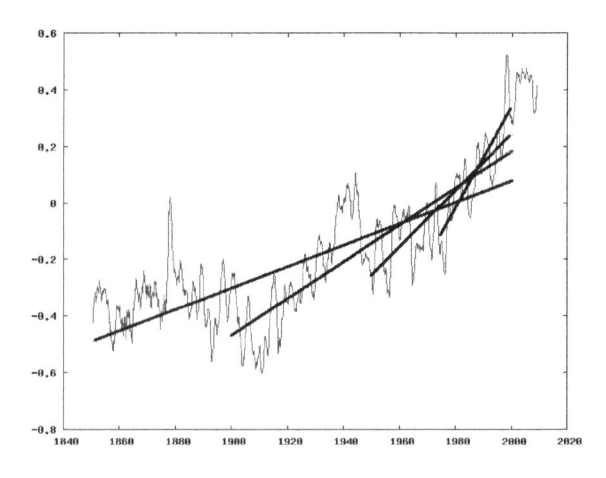

Figure 21 – Plotting of 150-year, 100-year, 50-year, and 25-year trends on variance adjusted HadCRUT3v data.

THE DECLINE

Looking at the graph above, we can see that towards the end a very large spike in temperatures occurs followed by what looks like a decline. The spike in temperatures during 1998 coincides with a very strong El Nino event. So now we will do the same as the IPCC and plot a trend line for the last nine years on the same data set.

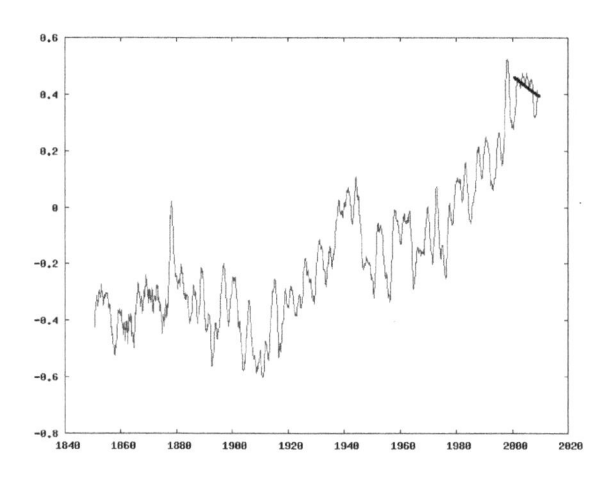

Figure 22 – Plotting a 9-year trend line on variance adjusted HadCRUT3v data.

Please notice that the trend line for 2001 – 2010 has a negative slope. This indicates that cooling took place for those nine years, despite increasing levels of carbon dioxide.

As an aside, please notice the Y (vertical) axis of these graphs. They are spaced so that this looks like a significant amount of warming, but the scale is in tenths (1/10) of a degree Celsius. So the graph indicates a rise of less than one degree Celsius in the last 160 years, which is well below the predictions of global warming alarmists.

Below we can show this cooling trend line existing in both the HadCRUT3v variance adjusted data (from 2001 - 2010) and the GISTEMP data (from 2002 - 2010). Both of these are data sets that have been highly molested in order to show the most possible global warming.

As previously mentioned, the HadCRUT3v data is compiled by the Hadley Center and by the Climatic Research Unit at East Anglia University. The GISTEMP data is compiled by NASA's Goddard Institute for Space Studies. It is oddly humorous that they can adjust the data to show ridiculous amounts of warming, but they can't get rid of the cooling since 2001.

Please note, again, that these two lines, both showing recent cooling trends, are plotted using the exact same technique that the IPCC used in the initial graphic of this chapter. Keep in mind that this demonstrates nearly a decade of cooling despite rising levels of carbon dioxide. This implies poor (or negative) correlation of temperature and carbon dioxide for nearly a decade. We have zoomed in on the time scale to show the most recent decade of global mean temperatures. In no way does this affect the plotting process of generating a trend line.

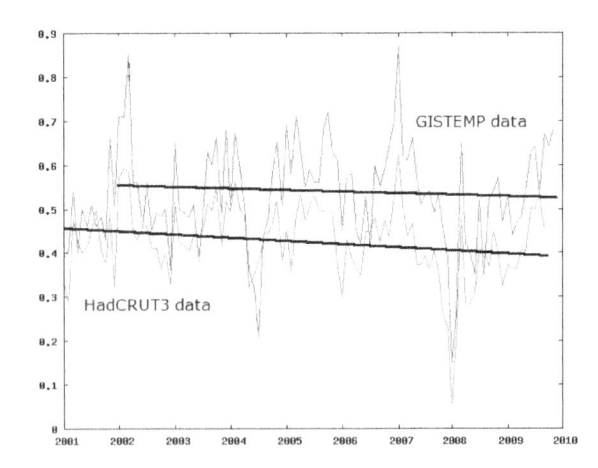

Figure 23 – HadCRUT3v 9 year trend line, and GISTEMP 8 year trend line. Both show almost a decade of cooling.

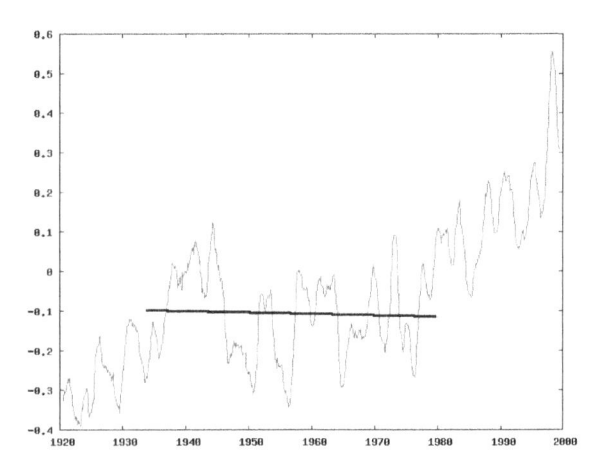

Figure 24 – HadCRUT3v 46-year trend line showing cooling from 1934 – 1980.

EVEN MORE COOLING

Note that we can also demonstrate the well-known forty-year cooling period from 1940 - 1980 in both the HadCRUT3v data and the GISTEMP data using this method. This cooling period is another thorn in the side of the alarmist camp. Using this technique you can actually extend the cooling period six years further to 1934 - 1980, which means 46 years of cooling, despite rising carbon dioxide levels.

If the IPCC expects us to accept their careful placing of trend lines to show accelerating anthropogenic global warming then we necessarily must accept that there has been at the very least 8 to 9 years of recent cooling.

BEHIND CLOSED DOORS

Don't think that this is merely a statistical manipulation of numbers to give the impression that the globe is cooling. Global warming has been missing for nearly a decade, despite rising levels of carbon dioxide. Fortunately, even the scientists implicated in ClimateGate have noticed the absence of global warming for the last decade. Kevin Trenberth, a lead author of the 1995, 2001, and 2007 IPCC Scientific Assessment of Climate Change, said it best:

"Hi all,

Well I have my own article on where the heck is global warming? We are asking that here in Boulder where we have broken records the past two days for the coldest days on record. We had 4 inches of snow. The high the last 2 days was below 30F and the normal is 69F, and it smashed the previous records for these days by 10F. The low was about 18F and also a record low, well below the previous record low. This is January weather (see the Rockies baseball playoff game was canceled on Saturday and then played last night in below freezing weather).

The fact is that we can't account for the lack of warming at the moment and it is a travesty that we can't. The CERES data published in the August BAMS 09 supplement on 2008 shows there should be even more warming: but the data are surely wrong. Our observing system is inadequate."

- Kevin Trenberth[39]

Privately, we see that Phil Jones, the head of the East Anglia University Climatic Research Unit, agrees with the fact that we've had cooling for nearly a decade:

"John Christy,

The scientific community would come down on me in no uncertain terms if I said the world had cooled from 1998. OK it has but it is only 7 years of data and it isn't statistically significant."

- Phil Jones[40] *in 2005*

Well, thank you for admitting that, Phil. Since the trend has continued, by your own words we have had twelve years of cooling (at least as of February 2010, the time of this writing.) But why didn't Phil Jones come out and tell the truth? Jones admits that the "scientific community" would "come down on him" should he reveal the fact that the globe has been cooling. Heaven forbid a scientist tell the truth. Keep in mind that Jones is the head of the CRU. Imagine how much pressure a junior scientist must feel if they want to keep their job?

As shown elsewhere in this book, the core group of ClimateGate scientists were adamant about preventing any anti-global warming literature from being published. Should Phil Jones ever admit that global warming has gone missing he would most likely never receive research money again, nor would ever be published again. He would likely lose his spots on the review panels of prestigious journals and his career would effectively be over.

It's also important to see that Jones, despite being privately aware of the global cooling trend, publicly maintained that anthropogenic global warming was a major threat to man and kept advocating spending billions of dollars on research. Despite his scientific arrogance, many people consider the billions of dollars unnecessarily spent on global warming research "statistically significant".

Further, there is another ClimateGate email exchange between Mick Kelly and Phil Jones:

"Phil Jones,

Just updated my global temperature trend graphic for a public talk and noted that the level has really been quite stable since 2000 or so and 2008 doesn't look too hot.

Anticipating the skeptics latching on to this soon, if they haven't done already, has anyone had a good look at the large-scale circulation anomalies over this period? I haven't noticed anything consistent coming up in the annual climate reviews but then I wasn't really looking.

Be awkward if we went through a early 1940s type swing!"

- Mick Kelly

To which Phil Jones replies:

"Mick Kelly,

They have noticed for years - mostly with respect to the warm year of 1998. The recent coolish years down to La Nina. When I get this question I have 1991-2000 and 2001-2007/8 averages to hand. Last time I did this they were about 0.2 different, which is what you'd expect."

- Phil Jones

Notice that Jones has a "stock answer" prepared to try to throw global warming skeptics off, instead of just coming out and admitting that we have been experiencing cooling since 1998. Kelly responds:

"Phil Jones,

Yeah, it wasn't so much 1998 and all that that I was concerned about, used to dealing with that, but the possibility that we might be going through a longer - 10 year - period of relatively stable temperatures beyond what you might expect from La Nina etc.

Speculation, but if I see this as a possibility then others might also.

Anyway, I'll maybe cut the last few points off the filtered curve before I give the talk again as that's trending down as a result of the end effects and the recent cold-ish years.

Enjoy Iceland and pass on my best wishes to Astrid."

- Mick Kelly

Notice that Mick Kelly writes that he will "cut the last few points off the curve" before he gives another presentation on global warming. This appears to be very typical of alarmist scientists; to alter the data in order to make it look like global warming is occurring at some catastrophic rate and trying to use this to influence people. It is pure deceit. And by his nonchalant admission of this to Jones, we can infer that they are both well aware of this mechanism of deceit. Apparently, it is no big deal to lie and misrepresent the data. It seems to be the rule instead of the exception.

Again, this shows that both Mick Kelly and Phil Jones are fully aware of the cooling since 1998, but still carry-on pushing the global warming message.

If it's not enough to show that we have had cooling for the last decade by using the same procedures as the IPCC and by hearing it directly from the horse's mouth, then we can consider the coup-de-grace to be the statistical analysis by the physicist Luboš Motl.

Dr. Motl used the UAH satellite temperature data, which is widely regarded as the most accurate data for our global mean temperature (please see Chapter 9) and mathematically proved that there has been no statistically significant global warming since 1995.

After calculating the slope of the mean temperatures from 1995 to 2009 he successfully demonstrated that it is only "somewhat more likely than not" that we have experienced global warming since 1995. This phrasing implies that any warming that may have occurred since 1995 is not statistically significant.

This should be more than enough evidence to demonstrate that we have indeed been experiencing global cooling for at least a decade.

Chapter 7

Ice Melts

"These figures are fresh. Some of the models say that there is a 75 per cent chance that the entire north polar ice cap, during the summer months, could be completely ice-free within five to seven years."

– Al Gore, at the Copenhagen Climate Change Summit, December 2009 [The scientist he quoted refuted that claim the same day]

This just in: ice melts. We've known since we were little children that when it is warmer ice will melt, and that when it is colder ice will form. Things could not be any simpler, and yet this natural phenomenon is the cause for so much alarmist hysteria.

In this chapter you will see how our recent ability to quantitatively measure sea ice has become a constant reason for certain people to become alarmed.

ARCTIC ICE

Alarmists frequently claim that Arctic ice is melting at increasing rates. What they don't tell you is that while it is true Arctic sea ice melts during the summer, it also forms during the winter. This is a natural cycle. In fact, when I look at the National Snow and Ice Data Center's (NSIDC) website, I see that it says Arctic ice is currently (December 2009) forming at a rate of 26,000 square miles per day. I guess those are headlines that alarmists don't like to read.

It is true that during this most recent warm period Arctic sea ice has decreased in overall quantity. This is to be expected when we are experiencing warming. In fact we will look at some historical evidence of Arctic sea ice severely melting during the 1910 – 1940 warming period, only to be reformed during the 1940 – 1975 cooling period.

We will now look at several graphics from the NSIDC to help us understand the seasonal nature of Arctic sea ice melting and forming.

You'll have to pay particularly careful attention to this graph as the translation to grayscale makes it difficult to read. The top two lines are the 1979 – 2000 average Arctic sea ice extent and 1979 – 2009 average Arctic sea ice extent, respectively. The next, thinner, line is the Arctic sea ice extent for 2009, and the dashed line is the low year of 2007.

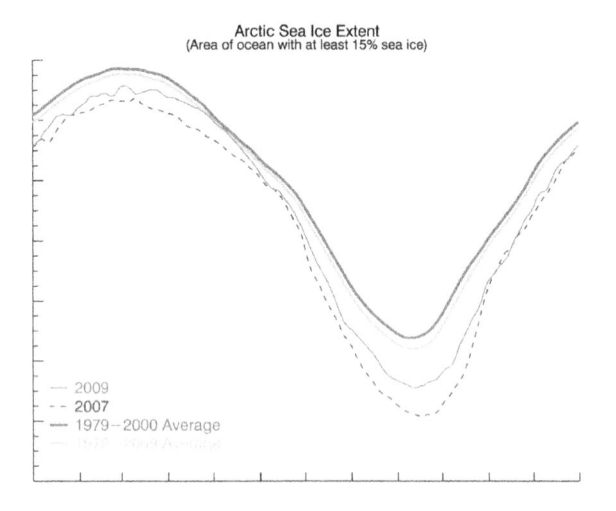

Figure 25 - Arctic sea ice extent, showing averages for 1979 - 2000, 1979 - 2009, and for 2009 and 2007. Courtesy of NSIDC.

Before we discuss the graph, keep in mind that this is only showing Arctic sea ice extent during our warming period (notice 1979 as a start date, the same as the start of the UAH satellite data – the Arctic sea ice extent is measured by satellite – this is right at the start of our current warm period.) Since we are looking at the quantity of ice during a warming period, it will necessarily display a decrease in Arctic sea ice.

First, when we look at the graph, we can see that Arctic sea ice is forming at the first part of the graph (this represents January.) Around April the ice extent starts plummeting until it reaches a minimum in September / October, and then afterwards it rapidly forms again. Sea ice forming and melting throughout the year is cyclical, natural, and expected.

Indeed, if you compare the average Arctic sea ice to the amount for 2009 you will see that 2009 has less overall sea ice. But, if you look at the graph, Arctic ice is up since 2007!

The conclusions we can logically draw from the data are: 1) Arctic sea ice forms and melts regularly throughout the year, 2) Arctic sea ice is down from the earlier average, which is to be expected because we are in a period of warmer temperatures, 3) Arctic sea ice is up since 2007, and 4) there is no reason to believe that Arctic sea ice is melting at increasing or alarming rates.

Alarmists make this gradually shrinking ice during a warm period appear unique.

The reality is that it is not unique. We just have only had a satellite looking at the ice quantity since we've been in a warm period. But satellite data is not the only data that exists relating to Arctic sea ice extent.

On November 2, 1922 the Washington Post ran a story titled "The Changing Arctic." Some of the text follows:

"The Arctic Ocean is warming up, icebergs are growing scarcer and in some places the seals are finding the water too hot, according to a report to the Commerce Department yesterday from Consul Ifft, at Bergen, Norway.

Reports from fishermen, seal hunters and explorers, he declared, all point to a radical change in climate conditions and hitherto unheard-of temperatures in the Arctic zone. Exploration expeditions report that scarcely any ice has been met with as far north as 81 degrees 29 minutes. Soundings to a depth of 3,100 meters showed the gulf stream still very warm.

Great masses of ice have been replaced by moraines of earth and stones, the report continued, while at many points well known glaciers have entirely disappeared. Very few seals and no white fish are found in the eastern Arctic, while vast shoals of herring and smelts, which have never before ventured so far north, are being encountered in the old seal fishing grounds."

I've included the full text of the article. Clearly, this demonstrates that Arctic sea ice melting has previously occurred in recent history. The Arctic sea ice that melted around 1922 reformed during the 1940 – 1975 cooling period.

Polar bears, penguins, and other Arctic animals were able to adapt to and have survived these natural cycles since the dawn of their existence.

November, 1922. MONTHLY WEATHER REVIEW.

THE CHANGING ARCTIC.

By George Nicolas Ifft.

[Under date of October 10, 1922, the American consul at Bergen, Norway, submitted the following report to the State Department, Washington, D. C.]

The Arctic seems to be warming up. Reports from fishermen, seal hunters, and explorers who sail the seas about Spitzbergen and the eastern Arctic, all point to a radical change in climatic conditions, and hitherto unheard-of high temperatures in that part of the earth's surface.

In August, 1922, the Norwegian Department of Commerce sent an expedition to Spitzbergen and Bear Island under the leadership of Dr. Adolf Hoel, lecturer on geology at the University of Christiania. Its purpose was to survey and chart the lands adjacent to the Norwegian mines on those islands, take soundings of the adjacent waters, and make other oceanographic investigations.

Dr. Hoel, who has just returned, reports the location of hitherto unknown coal deposits on the eastern shores of Advent Bay—deposits of vast extent and superior quality. This is regarded as of first importance, as so far most of the coal mined by the Norwegian companies on those islands has not been of the best quality.

R. L. Holmes: Quart. Journ. Royal Meteorol. Soc., January, 1905.

Figure 26 - Part 1/3 of "The Changing Arctic"

The oceanographic observations have, however, been even more interesting. Ice conditions were exceptional. In fact, so little ice has never before been noted. The expedition all but established a record, sailing as far north as 81° 29′ in ice-free water. This is the farthest north ever reached with modern oceanographic apparatus.

The character of the waters of the great polar basin has heretofore been practically unknown. Dr. Hoel reports that he made a section of the Gulf Stream at 81° north latitude and took soundings to a depth of 3,100 meters. These show the Gulf Stream very warm, and it could be traced as a surface current till beyond the 81st parallel. The warmth of the waters makes it probable that the favorable ice conditions will continue for some time.

Later a section was taken of the Gulf Stream off Bear Island and off the Isfjord, as well as a section of the cold current that comes down along the west coast of Spitzbergen off the south cape.

In connection with Dr. Hoel's report, it is of interest to note the unusually warm summer in Arctic Norway and the observations of Capt. Martin Ingebrigtsen, who has sailed the eastern Arctic for 54 years past. He says that he first noted warmer conditions in 1918, that since that time it has steadily gotten warmer, and that to-day the Arctic of that region is not recognizable as the same region of 1868 to 1917.

Figure 27 - Part 2/3 of "The Changing Arctic"

Many old landmarks are so changed as to be unrecognizable. Where formerly great masses of ice were found, there are now often moraines, accumulations of earth and stones. At many points where glaciers formerly extended far into the sea they have entirely disappeared.

The change in temperature, says Captain Ingebrigtsen, has also brought about great change in the flora and fauna of the Arctic. This summer he sought for white fish in Spitzbergen waters. Formerly great shoals of them were found there. This year he saw none, although he visited all the old fishing grounds.

There were few seal in Spitzbergen waters this year, the catch being far under the average. This, however, did not surprise the captain. He pointed out that formerly the waters about Spitzbergen held an even summer temperature of about 3° Celsius; this year recorded temperatures up to 15°, and last winter the ocean did not freeze over even on the north coast of Spitzbergen.

With the disappearance of white fish and seal has come other life in these waters. This year herring in great shoals were found along the west coast of Spitzbergen, all the way from the fry to the veritable great herring. Shoals of smelt were also met with.

Figure 28 - Part 3/3 of "The Changing Arctic"

ANTARCTIC ICE

On the opposite side of the globe rests Antarctica.[41] Antarctica is Earth's southernmost continent, underlying the South Pole. It is situated in the Antarctic region of the southern hemisphere, almost entirely south of the Antarctic Circle, and is surrounded by the Southern Ocean. About 98% of Antarctica's 5,400,000 square miles is covered by ice, which averages 1 mile in thickness.

Only cold-adapted plants and animals survive there, including penguins, seals, many types of algae and other microorganisms, and tundra vegetation.

Antarctica is the coldest place on Earth. The coldest natural temperature ever recorded on Earth was −89.2 °C (−128.6 °F) at the Russian Vostok Station in Antarctica on 21 July 1983. Antarctica is a vast frozen desert with little precipitation.

Keep in mind that we do not know much about the long-term history of Antarctica. It was officially spotted in 1820 by the Russian expedition of Mikhail Lazarev and Fabian Gottlieb von Bellingshausen. The continent, however, remained largely neglected for the rest of the 19th century because of its hostile environment, lack of resources, and isolation. This means all of our knowledge on the land size of Antarctica has been learned while we are exiting the Little Ice Age and experiencing warming.

The Wilkins Ice Shelf is a thick floating platform of ice located on the southwest side of the Antarctic Pennisula, primarily in the Wilkins Sound. This ice shelf is the cause for much alarm because in 2008 it appeared to be on the verge of collapsing off of the side of Antarctica. But then it froze back in place, and then in 2009 again it appeared to be on the verge of collapse. Apparently the breakup of this shelf is a regular, usual occurrence.

Figure 29 - Location of the Wilkins Ice Shelf in Antarctica. It is located in the circled region, on the southwest side of the Antarctic Peninsula.

In fact, the media seems to report this story almost annually. They use it as proof that humans are contributing to the catastrophic destruction of the planet. The story is such that the media reuses the same photos each year. How easy it must be to reload this story each year and just change the date on the article.

By Andrea Thompson
LiveScience
updated 12:20 p.m. PT, Tues., March. 25, 2008

A vast ice shelf hanging on by a thin strip looks to be the next chunk to break off from the Antarctic Peninsula, the latest sign of global warming's impact on Earth's southernmost continent.

Figure 30 - Story about the Wilkins ice shelf "on the verge of collapsing." From The MSNBC, March 25, 2008.

In 2008 this was the "latest sign of global warming's impact." The next year they used the same photograph, and almost the exact same story text. Talk about lazy reporting!

Antarctic ice shelf half the size of Scotland on verge of collapse

Paul Harris in New York
The Observer, Sunday 5 April 2009
Article history

This picture shows part of the Wilkins ice shelf as it began to break apart. Jim Elliott/British Antarctic Survey/AP

A huge ice shelf in the Antarctic is in the last stages of collapse and could break up within days in the latest sign of how global warming is thought to be changing the face of the planet

Figure 31 - Story about the Wilkins Ice Shelf "on the verge of collapsing." From The Observer, April 5, 2009.

National Geographic covered the story as well. Their article mentions, "It is an event we don't get to see very often." How rare can it be when it happened twice in a row?

Again, ice melting is consistent with a warming period. There is zero scientific evidence that supports the idea that this ice melting is occurring at any rate greater than that of natural variations. In other words, there is no evidence that this melting is related to anything done by man.

But, we need to stop focusing on one ice shelf way out on the Antarctic Peninsula. If you look at the graphic previously shown, Antarctica is huge.

Let's look at the sea ice extent trends for the whole of Antarctica. In the following graph you can see that since 1979 Antarctica sea ice is, on average growing.

Be sure to note the cyclical nature of the melting and growing phases of the Antarctic sea ice. Many different explanations are offered for the growing Antarctic ice, including the hole in the ozone layer and abnormal sea tides.

Ice core drilling off Australia's Davis Station in East Antarctica by the Antarctic Climate and Ecosystems Co-Operative Research Centre shows that last year, the ice had a maximum thickness of 1.89 meters, its densest in 10 years. The average thickness of the ice at Davis since the 1950s is 1.67 meters.[42]

The fact that Antarctic ice is growing and getting thicker does not support a picture of rapidly melting Antarctic ice, nor does it support any alarming claims that Antarctic ice is in danger of melting away.

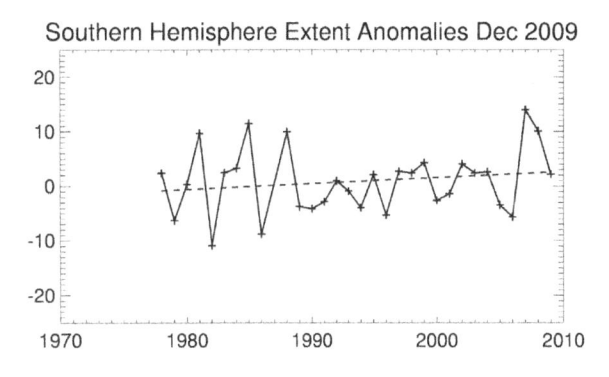

Figure 32 - Antarctic sea ice extent increasing since 1979. Courtesy of NSIDC.

In the next graph we will look at the cyclical nature of Antarctic ice melting and growing. You'll see that ice grows from about March to September, and melts the other parts of the year. Remember that Antarctica is part of the southern hemisphere, so it's seasons are opposite of what we would expect in the northern hemisphere. The cyclical nature of this is demonstrated by the large, sine wave looking part of the graph.

On the bottom part of the graph is another line that demonstrates the anomaly of sea ice extent from the 1979 – 2000 average.

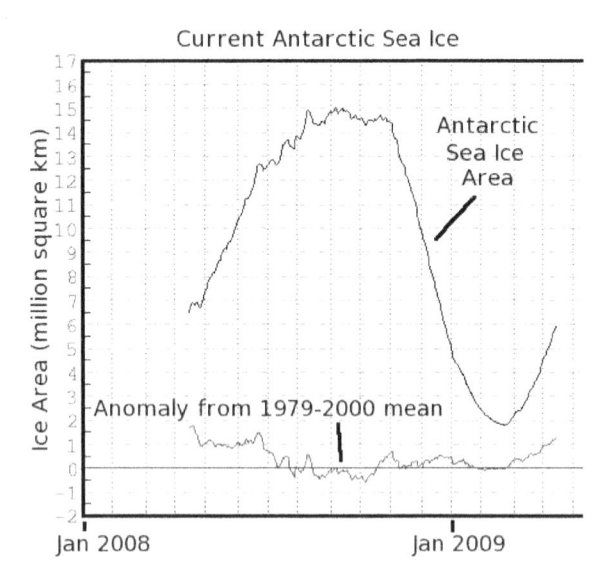

Figure 33 - Antarctic sea ice growth and melting detail.

The facts are that Arctic sea ice is up since 2007 and Antarctic sea ice is growing. This is does not support a picture of catastrophic anthropogenic global warming.

SEA LEVEL RISE

When ice melts it changes phase into water. In accordance with the alarmist claims of melting Arctic ice, Antarctic ice, and glaciers we get the predictions of sea levels rising at record rates that will devastate the world. We will briefly look the alarmist sea level rise predictions and show that they are not scientifically sound.

Sea level rise is primarily driven by two factors. The first is that increasing temperatures cause thermal expansion of water, and the second is the addition of water to the oceans from melting ice.

Approximately 20,000 years ago, during the Last Glacial Maximum, sea levels were at a minimum. This was a period of maximal ice extent that coincided with our most recent major ice age. Sea levels were at a minimum and glaciers and other ice sheets formed at a maximum. During this period of time, approximately 1/3rd of the earth was covered in ice. Theoretically, a person could walk from North America to Europe across the frozen north Atlantic ice sheet.

The ice that formed in the last major ice age has indeed been slowly melting over time, and this has caused a rise in sea levels. This is to be expected and it is not unusual. Water froze during the ice age, and since we left the last major ice age, the ice has been melting.

Ice Melts

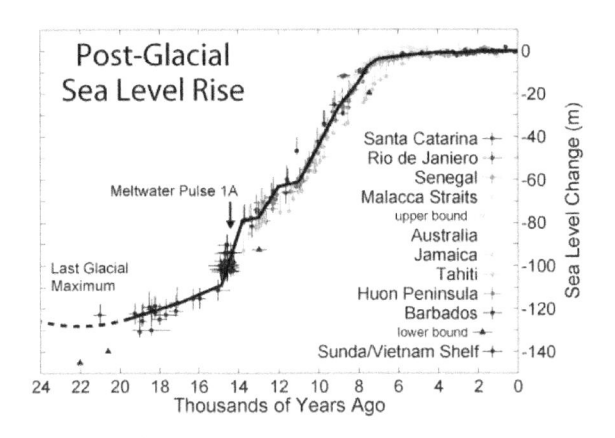

Figure 34 - Rate of sea level rise over the last 20,000 years.

As we can see in the above graph of sea level rise since the Last Glacial Maximum, sea level rise has not been at a constant rate. In the last 20,000 years sea level has risen by over 120 meters as a result of melting of major ice sheets due to natural forces. A rapid rise took place between 15,000 and 6,000 years ago at an average rate of 10 millimeters per year, which accounted for 90 meters (the majority) of the rise. It's important to note that this rise between 15,000 and 6,000 years ago was at a rate much higher than we have ever experienced in recent times and that this occurred without any human intervention.

Clearly we see that sea levels change through time. During colder periods sea levels drop, and during warmer periods sea levels rise. This has occurred regularly while the earth slumbered along for millions and millions of years.

In the 20th century sea level has risen at a rate of 1.8 millimeters per year (that is less than 2/10th's of a centimeter per year.) This rate of 1.8 millimeters per year has remained relatively constant since at least the 1850s (when humans started producing greenhouse gases.) If anything, since around 2004 the rate has slowed down. Below is a graphic of global mean sea level rise since 1880. Notice that it has remained relatively constant through this period of time.

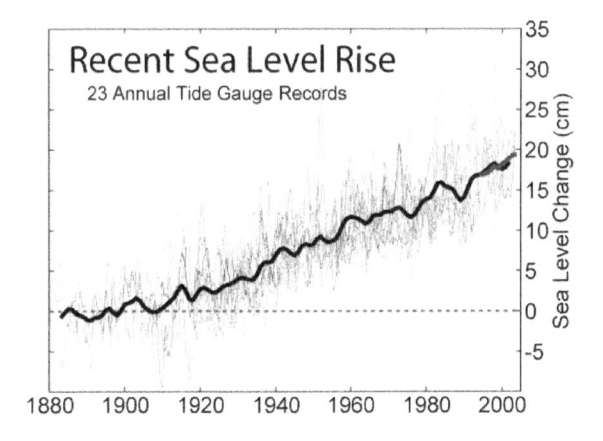

Figure 35 - Sea level rise since 1880. Sea levels have risen at a relatively constant rate.

Below is a graph of sea level rise since 1994. Notice that in recent times there has been a decrease in the rate of rise.

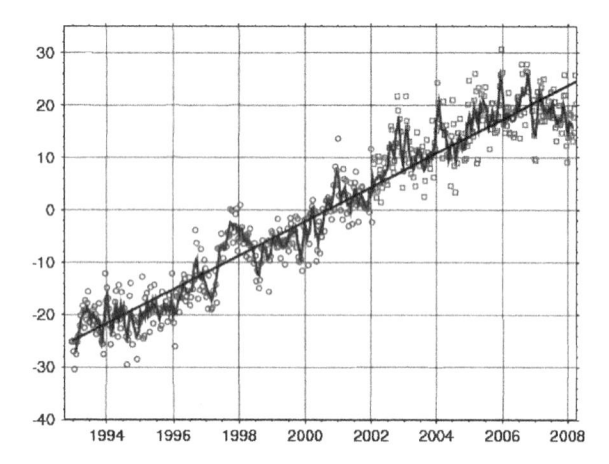

Figure 36 - Sea level rise since 1994. In recent times (starting around 2004) we see a decrease in the rate of sea level rise.

The Potsdam Institute for Climate Impact Research predicted that sea levels could rise by 6 feet by the year 2100. This prediction, despite however scary it may sound, is not based on science. It is based on assumption.

To make such a ludicrous prediction, they took the 7 inches of sea level rise in the past 120 years, which is a normal and not unusual rate, they then attributed it to a 0.7' Celsius rise in temperatures. The institute then assumed a rise in global temperatures of 6.4' Celsius by the year 2100, which is not based on science – temperatures are/were rising at the not alarming rate of 0.6' Celsius per century. So essentially they just made up a number and used it to instill fear in people.

Occasionally, NASA writes articles about the current state of the climate. Generally when we hear about NASA we associate them with everything good about science. They put a man on the moon! Unfortunately, their climate team is nothing more than a group of alarmists that have become known as the propaganda wing of this used-to-be-great institution. The following quote is from their most recent climate article:[43]

"[Antarctica] has been losing more than a hundred cubic kilometers (24 cubic miles) of ice each year since 2002" and that *"if all of this ice melted, it would raise global sea level by about 60 meters (197 feet)."*

Without a doubt, this sounds devastating. A 60-meter rise in sea levels would wipe out a good portion of the world's population. Unfortunately, there are two problems with this claim.

The first, as we previously saw, is that Antarctic sea ice extent is growing, not shrinking. Temperatures in the Antarctic are experiencing a cooling trend, not a warming one. And we saw that, if anything, the rate of sea level rise is decreasing.

The second is that Antarctica contains approximately 30,000,000 cubic kilometers of ice.[44] At NASA's claimed melting rate of 100 cubic kilometers per year, it would 300,000 years for the ice to melt and cause the sea level rises. Considering the rate of which major ice ages occur, NASA's threat could never occur. Indeed, NASA is playing the role of an alarmist. Clearly this isn't about the science, it is about the politics.

The fact that sea levels have remained relatively constant for the last 130 years, and, if anything, are slowing their rate of rise, is wholly inconsistent with the idea of catastrophic anthropogenic global warming. Simply put, there is no science that supports the idea that sea level rise rates will suddenly increase rapidly and wreck havoc upon the world.

Chapter 8

The Himalayan Glacier Lie

"It related to several countries in this region and their water sources. We thought that if we can highlight it, it will impact policy-makers and politicians and encourage them to take some concrete action."

– Murari Lal, admitting that the IPCC knowingly used fake data in it's most recent report.

Remember the International Panel on Climate Change (IPCC)? They are the group that writes huge assessment reports in order to convince legislators that catastrophic global warming is real and is a threat in order to persuade them to pass emissions curbing legislation.

In their Fourth Assessment Report they made the following statement:

"Glaciers in the Himalaya are receding faster than in any other part of the world (see Table 10.9) and, if the present rate continues, the likelihood of them disappearing by the year 2035 and perhaps sooner is very high if the Earth keeps warming at the current rate. Its total area will likely shrink from the present 500,000 to 100,000 square kilometers by the year 2035 (WWF, 2005.)" – IPCC AR4 WG2 Ch 10, p.493

That, of course, sounds terrible. Before we investigate this claim, it is important to remind ourselves that this document is supposed to be rock-solid. The IPCC prides itself by the number of authors and experts that supposedly contribute to and review this work.

So, let's dissect this. First, we see that this IPCC quote is referencing the WWF - The World Wildlife Foundation. The WWF does not write peer-reviewed literature. They are a foundation that writes opinion pieces in order to accomplish an agenda. So in order for the IPCC to quote them, the IPCC would have needed to refer to the source of the WWF article and verify the information. This did not happen. The original statement in the WWF article:

"In 1999, a report by the Working Group on Himalayan Glaciology (WGHG) of the International Commission for Snow and Ice (ICSI) stated: 'glaciers in the Himalayas are receding faster than in any other part of the world and, if the present rate continues, the likelihood of them disappearing by the year 2035 is very high.' Direct observation of a select few snout positions out of the thousands of Himalayan glaciers indicate that they have been in a general state of decline over, at least, the past 150 years. The prediction that 'glaciers in the region will vanish within 40 years as a result of global warming' and that the flow of Himalayan rivers will 'eventually diminish, resulting in widespread water shortages' (New Scientist 1999; 1999, 2003) is equally disturbing" – WWF 2005

Clearly, we can see that the WWF is quoting *New Scientist* in their article. So not only did the IPCC quote non peer-reviewed literature, they quoted something that was second-hand information from somewhere else. *New Scientist* is a magazine, not peer-reviewed literature.

This is the *New Scientist* article that was quoted by the WWF:

"A new study, due to be presented in July to the International Commission on Snow and Ice (ICSI), predicts that most of the glaciers in the region will vanish within 40 years as a result of global warming. 'All the glaciers in the middle Himalayas are retreating,' says Syed Hasnain of Jawaharlal Nehru university in Delhi, the chief author of the ICSI report...Hasnain's four-year study indicates that all the glaciers in the central and eastern Himalayas could disappear by 2035 at their present rate of decline...Hasnain's working group on Himalayan glaciology, set up by the ICSI, has found that glaciers are receding faster in the Himalayas than anywhere else on Earth. Hasnain warns that as the glaciers disappear, the flow of these rivers will become less reliable and eventually diminish, resulting in widespread water shortages." - New Scientist 1999

So this *New Scientist* article is actually nothing more than a paraphrasing of Syed Hasnian about the Working Group on Himalayan Glaciology of the International Commission on Snow and Ice report from 1999. Notice that what was paraphrasing in the *New Scientist* article has become direct quotes in the WWF report.

John Nielsen-Gammon[45] of the Houston Chronicle managed to dig up the 1999

The Himalayan Glacier Lie

WGHG/ICSI report and found that the report does not compare Himalayan glacier retreat with other rates of recession and does not mention a date for the disappearance of Himalayan glaciers. In other words, none of the statements made by the IPCC are supported in the original report. The statements have been fabricated.

To recap: the IPCC quoted the WWF. The WWF quoted *New Scientist*, and *New Scientist* paraphrased a conversation with an author about an upcoming report.

So where did this figure of Himalayan glacier disappearance by 2035 come from? Apparently the only source is the paraphrasing of Syed Hasnian. There is no science that supports a melting date of 2035.

Other literature (for example, Kotlyakov) regarding Himalayan glacier retreat reads as follows:

"The degradation of the extra-polar glaciation of the Earth will be apparent in rising ocean level already by the year 2050, and there will be a drastic rise of the ocean thereafter caused by the deglaciation-derived runoff. This period will last from 200 to 300 years. The extra-polar glaciation of the Earth will be decaying at rapid, catastrophic rates – its total area will shrink from 500,000 to 100,000 square kilometers by the year 2350. Glaciers will survive only in the mountains of inner Alaska, on some Arctic archipelagos, within Patagonian ice sheets, in the Karakoram Mountins, in the Himalayas, in some regions of Tibet and on the highest mountain peaks in the temperature latitudes." – Kotlyakov

Now, it is important to realize that Kotlyakov is alarmist rubbish too. But even in this piece he only predicts the "major loss" of the Himalayas by 2350, not 2035. Also, notice that he says the entire area of glaciation is 500,000 square kilometers; where-as the IPCC had the area of just the Himalayas as 500,000 square kilometers (it's really 33,000 square kilometers.)

So clearly, we can see that Hasnain's quotes (and thus the IPCC's chapter) on glacier recession are rubbish. Absolutely nothing in the world supports the IPCC claimed melting date of 2035 for the Himalayan glaciers. Continue reading, it gets worse.

As a result of some of the criticism the IPCC received for publishing such a ridiculous statement the Indian environment ministry released a report in November 2009 that concluded the Himalayan glaciers on the whole were retreating, but not at an alarming rate or any faster than glaciers on the rest of the globe. They specifically stated that the glacier retreat was not "historically alarming."

IPCC head, and fellow Indian, Rajendara Pachauri was upset that anyone would dare question the validity of the IPCC reports. He responds:

"We have a very clear idea of what is happening. I don't know why the minister is supporting this unsubstantiated research. It is an extremely arrogant statement."

He also called the report "voodoo science." This action is expected of someone who is grasping at straws. Why attack someone rather than actually look to see if there was an error?

As it turns out, Syed Hasnain denied ever making the statement that the Himalayan glaciers would melt by 2035. This means that the IPCC report, the WWF report, and the New Scientist report were all invalid and not based on anything. The primary source denied ever making the claim.

Now we must look at whether or not the IPCC was knowingly including bogus statements in their report.

Before[46] the 2007 report was published a leading Austrian glaciologist, Dr Georg Kaser, who was also a lead author on the 2007 report, brought the offending claim into light. He described Syed Hasnain's prediction of glaciers disappearing by 2035 as "so wrong that it is not even worth dismissing".

After it was found out that the IPCC published the report with this fake data in it, Dr. Murari Lal, a lead author of the IPCC chapter on glaciers, admitted[47] that the statement was only included in the report to put political pressure on world leaders. He openly admitted to putting fake data in the IPCC report. Murari Lal:

"It related to several countries in this region and their water sources. We thought that if we can highlight it, it will impact policy-makers and politicians and encourage them to take some concrete action ... It had importance for the region, so we thought we should put it in. ... We knew the WWF report with the 2035 date was "grey literature" [material not published in a peer-reviewed journal]. But it was never picked up by any of the authors in our working group, nor by any of the more than 500 external reviewers, by the governments to which it was sent, or by the final IPCC review editors ... We relied rather heavily on grey [non peer-reviewed] literature, including the WWF report."

Murari Lal goes on to blame the fact that the IPCC included literature they knew was bogus on Syed Hasnain's assertion of the 2035 date.

However, Syed Hasnain rejected the blame. In turn, he blamed the IPCC for misusing a remark he made to a journalist. "The magic number of 2035 has not [been] mentioned in any research papers written by me, as no peer-reviewed journal will accept speculative figures," he told *New Scientist*.[48]

So there we have it. The IPCC report knowingly included a bogus claim about melting Himalayan glaciers in order to put pressure on policy makers. The bogus claim has been picked up and spread by alarmists all over the world, including anti-scientists such as Al Gore. The claim has no basis in reality, none of the literature supports it, and the person who is claimed to have made the statement denies it.

It is an outright lie. Completely fake data.

The only comment the IPCC has made regarding this issue is that it apologizes for the statement being "poorly substantiated." It wasn't poorly substantiated; it was created out of thin air!

What did Rajendra Pachauri, the IPCC head, say after Murari Lal spilled the beans about knowingly including false data in the IPCC report? He initially stated that he had "absolutely no responsibility" for the falsified data, and that it was "the work of independent authors—they're responsible." Eventually he remarked that it was due to a serious system failure. What a stand up guy, right?

Pachauri's *coup-de-grace* is still to come. Pallava Bagla, a prominent science journalist, had told Pachauri that the leading glaciologists stated the Himalayan glacier melt date was wrong by at least 300 years. He told Pachauri this several months before the Copenhagen conference on climate in December 2009. This means that Rajendra Pachauri knew of the bad Himalayan glacier data at least 3 to 4 months before he accused people of being "arrogant" and using "voodoo science" for publicly pointing out that the data was wrong!

As far as glacial retreat is concerned, we must remember that ice melts. We are in a warming period so glaciers will melt. They melted during the 1910 – 1940 warming period, and then they grew again during the cooling period immediately after. A study by Huss, et al. states that glaciers were melting faster in the 1940s than they are today, despite our "higher temperatures", due to enhanced solar radiation.[49]

Lastly, we must point out the original WWF article stated that the rapidly melting glaciers would result in water shortages. We must understand that glacier melt is the process that provides water to nearly a billion people. If the glaciers were not melting there would be no water supply. So it is a contradiction to state that if glaciers melt water supplies will vanish because glaciers have to melt to get water in the first place. Alarmists can't have their cake and eat it too.

MORE BAD CITATIONS

If it is not enough that Rajendra Pachauri and the IPCC knowingly used fake data in their report in order to pressure policy makers, they have also littered their report with opinion pieces from advocacy groups.

In Chapter 13 of the IPCC report, the following statement is made:

"Up to 40% of the Amazonian forests could react drastically to even a slight reduction in precipitation; this means that the tropical vegetation, hydrology, and climate system in South America could change very rapidly to another steady state, not necessarily producing gradual changes between the current and future situation (Rowell and Moore, 2000). It is more probably that forests will be replaced by ecosystems that have more resistance to multiple stresses caused by temperature increase, droughts and fires, such as tropical savannas."

Again, that sure sounds alarming! Blogger Richard North[50] followed the sources, and found that Rowell and Moore is actually a WWF report. The actual statement written in the Rowell and Moore report reads as follows:

"Up to 40% of the Brazilian forest is extremely sensitive to small reductions in the amount of rainfall. In the 1998 dry season, some 270,000 sq. km of forest became vulnerable to fire, due to completely depleted plant-available water stored in the upper five meters of soil. A further 360,000 sq. km of forest had only 250 mm of plant-available soil water left."

And, in turn, Rowell and Moore reference an article by D. C. Nepstad in Nature[51]. The statement there reads as follows:

"Although logging and forest surface fires usually do not kill all trees, they severely damage forests. Logging companies in Amazonia

kill or damage 10-40% of the living biomass of forests through the harvest process. Logging also increases forest flammability by reducing forest leaf canopy coverage by 14-50%, allowing sunlight to penetrate to the forest floor, where it dries out the organic debris created by the logging."

So, working from the original Nature article all the way to the IPCC report, we see that "Logging companies in Amazonia kill or damage 10-40% of the living biomass of forests through the harvest process," magically turns into "up to 40% of the Brazilian forest is extremely sensitive to small reductions in the amount of rainfall," which then mutates into "up to 40% of the Amazonian forests could react drastically to even a slight reduction in precipitation".

Further, North researched the backgrounds of Rowell and Moore, and found Rowell to be a freelance journalist and an activist, and Moore to be a specialist in policy.

So, the IPCC based their prediction for the Amazon rain forest off of a second hand source written by two non-scientists who misquoted their original source. Clearly, the IPCC prediction for the fate of the Amazon rainforest is fabricated and there is no science backing it up. It is a lie.

Blogger Donna Laframboise of NOconsensus.org highlighted dozens of World Wildlife Foundation (WWF) citations in the latest IPCC report. Remember, the WWF is an agenda-driven organization and its reports are not peer-reviewed science. They are merely opinion pieces and nothing more. They hold as much scientific weight as a book report that you wrote in gradeschool.

For example[52], Mrs. Laframboise found that WWF reports were the only sources cited by the IPCC as proof regarding coastal development problems in Latin America, and they were quoted to define what the global average per capita "ecological footprint" is compared to the ecological footprint of central and eastern Europe. Also, when discussing disasters related to melting glaciers the IPCC quoted a WWF document and a never-published paper from 2002.

She found that the IPCC's declaration that "Changes in climate are affecting many mountain glaciers, with rapid glacier retreat documented in the Himalayas, Greenland, the European Alps, the Andes Cordillera and East Africa" is solely backed by a WWF report. Furthermore, when discussing coral reefs and mangroves, a WWF report is the only thing backing up the IPCC's statement that in "the Mesoamerican reef there are up to 25 times more fish of some species on reefs close to mangrove areas than in areas where mangroves have been destroyed."

Furthermore, Donna found that the IPCC relied upon Greenpeace reports to make alarming claims. Greenpeace is a known agenda-driven activist organization. These are pretty much the last people you would want to quote in a scientific article.

She found[53] the IPCC using Greenpeace reports to link climate change to coral reef degradation, to establish the lower-end of an estimate involving solar power plants (you can imagine Greenpeace easily overestimating the production capability of solar power plants!), and it also used Greenpeace reports to document where the "main wind-energy investments" are located globally.

Lastly, she points out that the authors of these opinion piece reports are not scientists. They are members of activist groups, such as WWF, Greenpeace, Friends of the Earth, Climate Action Network, and Environmental Defense. Why on earth are these people being quoted in a scientific document that is designed to persuade policy makers to believe that humans are causing catastrophic global warming? Sadly, I think we all know the answer to that question.

Richard Gray and Rebecca Lefort reported[54] about the IPCC using unscientific sources to backup their claims that observed reductions in mountain ice in the Andes, Alps and Africa were being caused by global warming. The IPCC cited two sources for this statement.

It turns out that the first source quoted by the IPCC was from an article published in [a rock] *Climbing* magazine, which was based on anecdotal evidence from mountaineers about the changes they were witnessing on the mountainsides around them.

The other source was an essay from a student studying for the equivalent of a master's degree at the University of Berne in Switzerland. As it turns out, the IPCC has quoted 9 different masters theses, and 31 different PhD theses. Obviously, since none of these are peer-reviewed, they are not valid sources of information, especially for a report that is designed to affect the outcomes of billions lives.

Blogger ClimateQuotes[55] found that the IPCC cited a boot-cleaning guide for Antarctic tour guide operators to support the idea that climate change has forced more stringent clothing decontamination

The Himalayan Glacier Lie

guidelines. Sadly, the boot-cleaning guide did not mention climate change.

Worse, the IPCC has quoted the New York Times as peer-reviewed literature to support the idea that unreliable electric power amplifies health concerns.

As I write this, many of the top climate scientists are referring to the IPCC with scorn. Andrew Weaver, one of Canada's leading climate scientists, says that the IPCC has dangerously crossed the line between climate science and climate advocacy.

The U.K. Government's Chief Scientist, John Beddington, has publicly called for more honesty in climate science. He openly disparaged dishonest scientists for exaggerating the predictions of catastrophic global warming.

And this is only the beginning. This admission of the IPCC using fake data happened merely days ago at the time of this writing. Time will tell how this affects things in the long run.

In the middle of this mess, the IPCC head Rajendra Pachauri made a rather timely announcement. He released a smutty romance novel named "Return to Almora." Apparently it is the story of an academic, a climate scientist perhaps, that flies around the world and has sex with women while referencing the Kamasutra. Sincerely, Pachauri is the laughing stock of the climate science world.

Figure 37 – The Love Guru, Rajendra Pachauri, IPCC head, with his smut novel "Return to Almora."

Chapter 10

The Quality Of U.S. Temperature Data

"Global Warming is a classic case of a belief system in cognitive dissonance with the data, and try as hard as you might, you can't convince a believer of anything, for their belief isn't based on evidence, but on a deep seated need to believe."

- Louis Hissink, paraphrasing Carl Sagan

Measuring the average temperature of the globe is a difficult process. It requires taking measurements during the morning and in the evening in nearly every part of the world. Unfortunately a lot of the collection is performed manually by humans and is therefore prone to error. The quality of global mean temperature data has been debated and questioned for many years. In this chapter we will look at the methods of collecting and processing U.S. temperature data.

Figure 38 – Courtesy of Anthony Watts, Dr. Roger Pielke Sr., and SurfaceStations.org

The above photo is from a temperature surface station in Hopkinsville, KY. See the white birdhouse looking thing hanging off the building? That's the thermometer. Initially this appears innocent, but notice that it comes off of the chimney of a brick house. Also, it's just above a black asphalt driveway, the big white thing is an air conditioning compressor, and it's directly over a Weber grill. All of these objects produce heat.

The data from this thermometer becomes a part of the US Historical Climatological Network (USHCN) Station of Record for Hopkinsville, KY. It would come as no surprise then that this station reports unusually high temperatures. This may initially, seem rather humorous as they are literally "cooking the books" at this station. But then it sinks in that this station data is being compiled in with the rest of the data from the country and eventually legislation is written based on the implications of this "cooked" (burnt?) data.

You may be wondering if this station is an outlier. Sadly, it is not. Over the next few pages we will detail different temperature measuring stations from around the United States that all contribute data to the U.S. Historical Climatological Network. To the best of our knowledge we will point out sources of heat contamination.

The Quality Of U.S. Temperature Data

Figure 39 – Courtesy of Anthony Watts, Dr. Roger Pielke Sr., and SurfaceStations.org

In this station we see the thermometer housing is placed directly next to a cement building, and directly next to an air conditioning compressor. Cement and brick buildings absorb heat all day long and radiate that heat during the evenings. This means that the station would give abnormally high day and night readings. This is a poorly sited surface station and will give poor temperature readings.

Figure 40 – Courtesy of Anthony Watts and SurfaceStations.org

In this station from Lovelock, Nevada, we see the thermometer housing is placed in the middle of an airport runway. There are multiple air conditioning compressors near the thermometer, and it is sitting on an asphalt runway. Also, airplanes are stored and parked directly next to the instrument as well. I wonder how much heat a turbine engine puts out. There was also found to be an incandescent light bulb in the thermometer shelter.

Figure 41 – Courtesy of Anthony Watts and SurfaceStations.org

In this station from Forest Grove, Oregon, we see the thermometer housing is placed directly next to a building, by a huge chain link fence, and directly next to an air conditioning compressor.

Figure 42 – Courtesy of Anthony Watts and SurfaceStations.org

In this station from Marysville, California, we see the thermometer housing is placed in an asphalt parking lot, with multiple air conditioner exhaust fans close by, and a cell tower in the middle of the parking lot, along with vehicle being parked there all day.

Figure 43 – Courtesy of Anthony Watts and SurfaceStations.org

In this station from Marysville, California we see the thermometer housing is placed directly next to two air conditioning units, is placed on top of a concrete patio, a galvanized steel cellular tower and a barbeque pit.

The Quality Of U.S. Temperature Data

81

Figure 44 – Courtesy of Anthony Watts, Dr. Roger Pielke Sr., and SurfaceStations.org

In this station from Tahoe City, California we see the thermometer housing is placed in close proximity to a tennis court, and directly next to a trash-burning barrel. Can you imagine the readings this thermometer gives when this station is burning trash? They may as well throw their readings in the trash too.

Figure 45 – Courtesy of Anthony Watts, Dr. Roger Pielke Sr., and SurfaceStations.org

In this especially horrible station from Roseburg, Oregon, we see the thermometer housing is placed on a dark rooftop. It is next to all sorts of satellites and other communication devices. Also, it is next to air conditioning units and a parking lot.

Figure 46 – Courtesy of Anthony Watts, Dr. Roger Pielke Sr., and SurfaceStations.org

In this station from Aberdeen, Washington, we see the thermometer housing is placed directly next to an asphalt parking lot. Imagine this thermometer sitting next to the asphalt all day and all night long, while the parking lot is full of cars, all collecting and radiating heat. Consider the warm exhaust from the vehicles affecting the thermometer readings.

Figure 47 – Courtesy of Anthony Watts, Dr. Roger Pielke Sr., and SurfaceStations.org

In this station from Lodi, California we see the thermometer housing is placed in an asphalt parking lot with barbeque pits and parking nearby. There's also a large steel trashcan next to the thermometer.

Keep in mind that this is just a random selection of stations. In a little while we will see that 90% of the stations are incapable of giving quality data.

Again, this may all seem hilarious, but we must realize that the data from these stations is being used to convince legislators to pass laws that will, literally, tax almost every thing that we do. Not quite as funny any more, is it?

If it seems like we are being a little too picky about the quality of the data of these surface stations, remember that we are measuring tiny changes in global mean temperature changes on the order of 0.6' Celsius in the last one hundred years.

Anthony Watts, author of one of the top science blogs on the Internet, has released a report titled "Is the U.S. Surface Temperature Record Reliable?"[56] I will give a brief overview of his report, and then cover additional items.

The official record of temperatures in the continental United States comes from a network of 1,221 climate-monitoring stations overseen by the National Weather Service, a department of the National Oceanic and Atmospheric Administration (NOAA). These surface stations are supposed to be setup in accordance with guidelines[57] given by the NOAA's Climate Reference Network (CRN).

Sites are rated on a scale of CRN 1 to CRN 5, with CRN 1 being a very well sited station, and CRN 5 being a very poorly sited station. Stations are sited to avoid problems with temperature and humidity (e.g., artificial heat sources, reflective surfaces, large bodies of water, being on a roof top, and being positioned on a parking lot or anywhere around concrete.), problems with precipitation (e.g., having tall objects close by), problems with solar radiation (e.g., being close to a building or other object that will block the sun), and lastly problems with wind.

Despite using the temperature data from these stations to try to enforce the idea of anthropogenic global warming, the NOAA has never actually surveyed the sites to see if their data is reliable. They merely setup guidelines to rate the sites, but have never actually rated the sites themselves. They leave it up to the individual surface station owner to try their best to follow the guidelines.

In 2007 a project was initiated by Watts to visually inspect and document the quality of the surface stations. So what did he find? Some of the stations that he found are pictured above. We can see the thermometers located next to asphalt parking lots, located next to air-conditioner exhausts, located on brick buildings, located next to BBQ pits, located next to trash burning barrels, located on rooftops, located near waste treatment facilities, and near sidewalks. All of these locations absorb and radiate heat. These all influence the thermometer to read higher than it would otherwise.

Let's look at the numbers.

There are 1,221 active climate-monitoring surface stations in the USA. Of those 1,221 stations, 948 have been evaluated by Watt's surface stations project. Of these 948, 90% were found to have been sited very poorly (CRN ratings of 3, 4, or 5.) Poorly sited means that they are most likely reporting higher temperatures than they should be. This means the overwhelming majority of stations in the USA have been reporting bad data. Please note that the USA data is generally considered the best data in the world. Yikes!

Here is a graphical breakdown of the sites and their quality rating as described by the NOAA CRN:

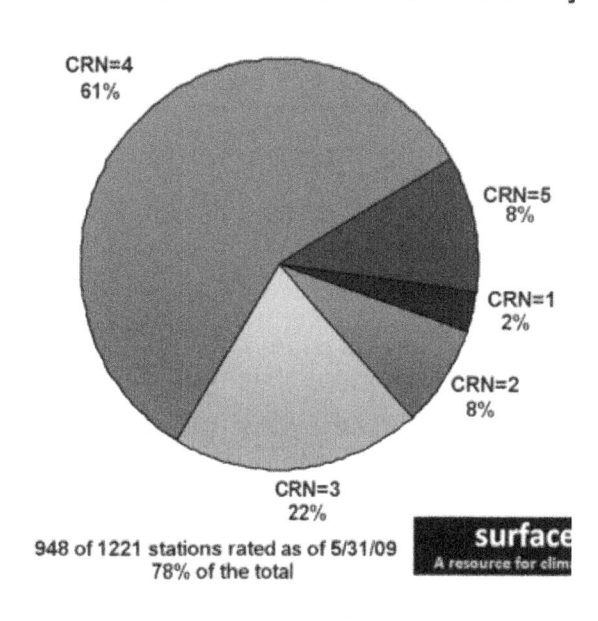

Figure 48 – Breakdown of the surface stations and their quality rating. Courtesy of Anthony Watts and surfacestations.org

Please notice that approximate temperature errors (as dictated by the NOAA CRN) are given for each CRN rating. The best sited stations, CRN 1 and 2, have an error less than 1.0′ Celsius. The poorly sited stations, CRN 3, 4, and 5, have an error of greater than or equal to 1.0′ Celsius.

As you can see, 90% of the surveyed surface stations have an error greater than or equal to 1.0' Celsius. This amount of error is greater than the amount of global warming we are supposed to have experienced from 1900 - 2000. Let me say that again. The amount of error in the data is greater than the amount of warming we are supposed to have undergone in the last 100 years.

The other 10% of the stations have been surveyed to show that they are capable of giving good data. It would be easy to assume that the other 10% give good data all the time, but unfortunately that is not the case. Let's see why.

To collect surface station data site observers are given a card on which they are supposed to fill out the maximum and minimum temperatures each day. At the end of the month they are supposed to submit their card to the National Climatic Data Center so that they can be compiled into the national database.

Well, what happens when a site observer doesn't work on a particular day? No readings occur. What happens when that observer is off because of a holiday? No readings occur. In fact, in Marysville, California, at Chico University Mr. Watts found that the temperature form for July 2007 had only 14 of 31 days completed. That is less than half a month's worth of data.

Missing data is not a rare phenomenon. Many sites have missing data. So even if a site is rated as capable of providing good data (CRN 1 or 2), the data still must be actually be recorded by someone. If that person is missing half the month, then you're not getting a satisfactory quantity of good data.

So what happens when stations have missing data? From Watts' report:

"[There exists] a data algorithm used by NCDC called FILNET, short for Fill Missing Original Data in the Network, that is used to "infill" missing data using interpolations of data from surrounding stations. After reading about it, I came to the conclusion that the NCDC uses FILNET to create "missing" data where none was ever actually measured."

The Quality Of U.S. Temperature Data

From a government report on the matter:

"Estimates for missing data are provided using a procedure similar to that used in SHAP [Station History Adjustment Program]. This adjustment uses the debiased data from the SHAP and fills in missing original data when needed (i.e. calculates estimated data) based on a "network" of the best correlated nearby stations. The FILNET program also completed the data adjustment process for stations that moved too often for SHAP to estimate the adjustments needed to de-bias the data."

Basically, if data is missing from a certain site then data is taken from near-by sites and adjusted to fit the missing site. Wow.

We see that 90% of the United States surface stations are not capable of providing quality data, and then we see that sites that have missing data have data filled in from surrounding sites by the FILNET program.

This means that a site which is capable of producing good data, if data is not read for a day, could have bad data filled in for it by this FILNET program. This means that even the 10% of sites that are capable of providing good data are contaminated by the 90% of sites that have bad data.

Now that we have shown that the quality of data from the stations is overwhelmingly poor, let's look a little more at the numbers and types of stations.

There are two main types of surface stations. There are stations that are found in suburban and urban areas, which are more likely to be contaminated by local heat sources, and then there are stations in rural areas, which are less likely to be contaminated by local heat sources.

Below is a great example of a well-sited rural station in Tucumcari, NM. Notice that there are no asphalt parking lots near-by, no airports with huge airplanes, no barbeque pits, and no trash burning barrels near-by. Notice that there are no large cement buildings that will radiate heat on it all day, nor are there big trees or large pieces of communications equipment next to it.

Figure 49 – Courtesy of Anthony Watts, Dr. Roger Pielke Sr., and SurfaceStations.org

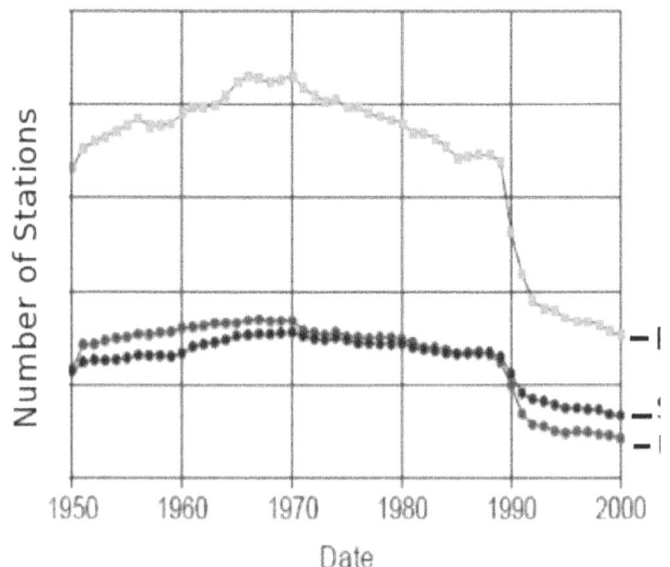

Figure 50 - Courtesy of ICECAP

In April 1978 there were over 6,000 surface stations and now we have dropped down to around 1,200. This means approximately 4,800 surface stations dropped out of the program and are no longer monitoring temperatures (and yet somehow some of these stations are still being represented in the data sets!) The vast majority of the stations that dropped out were rural stations. This is unfortunate because the rural stations are the ones that tend to give the most accurate data due to less urban heat contamination.

Here is a graphic displaying the relative surface station dropout of rural versus urban locations:

Because of the fact that the majority of stations which dropped out were rural ones, we can deduce that when the FILNET program is applied to stations that are missing data, data is more likely to come from urban (contaminated) locations than rural ones.

By now, I'm sure you are wondering if this can get any worse. Unfortunately it does.

After all the data is collected from the stations, and after missing station data is created by the FILNET program, we find that adjustments are applied to "homogenize" the data (that is, data is compared to surrounding stations and adjusted accordingly) that impart an even larger false warming trend.

Recently, emails secured by a Freedom of Information Act that was sent to NASA demonstrate the ridiculous lengths alarmist scientists go through to increase the amount of warming that we have seen in recent times. For example, 1934's mean temperature anomaly was adjusted 16% downwards from 1.459' Celsius to 1.227' Celsius, while 1998's mean temperature anomaly was adjusted 35% higher from 0.918' Celsius to 1.242' Celsius.

If you look at these numbers, you see that these alterations serve to make 1934 no longer be the warmest year ever, and they serve to make 1998 the warmest. The email that demonstrates this altering of temperatures is reproduced at the end of the chapter.

Below is an example of the "homogenized" data. The line with a flatter slope is the "raw" (unaltered) data. The line with a steeper slope is the "homogenized" (altered) data. You can clearly see the data from 1880 has been altered to appear cooler in the beginning, and warmer as time travels toward the present.

This does two things:

1) It makes it appear the past was cooler, which makes our current warming look unique, and

2) It gives the data a steeper warming trend, which makes our current warming look more severe.

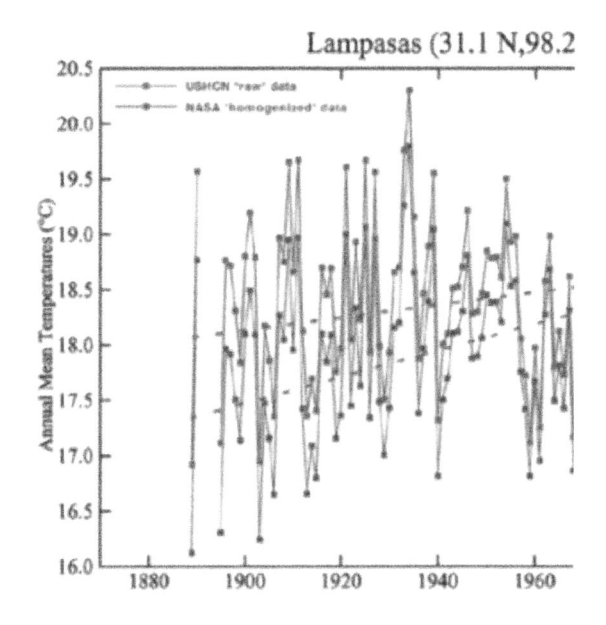

Figure 51 - Courtesy of Anthony Watts and SufaceStations.org. Raw vs. Homogenized Data. The steeper slope is the homogenized data. The other is the raw, unaltered data. Clearly, a warming bias has been introduced.

On the next page we can look at a graph released by the NOAA that demonstrates the differences between the raw and homogenized data sets. Note that this is a graph of differences; it shows the difference between the raw and homogenized data sets. Since this graph trends upwards, it means that the homogenized data sets have been reading much warmer than the raw data sets.

It's funny how this curve is very similar to the curve we see showing a warming trend.

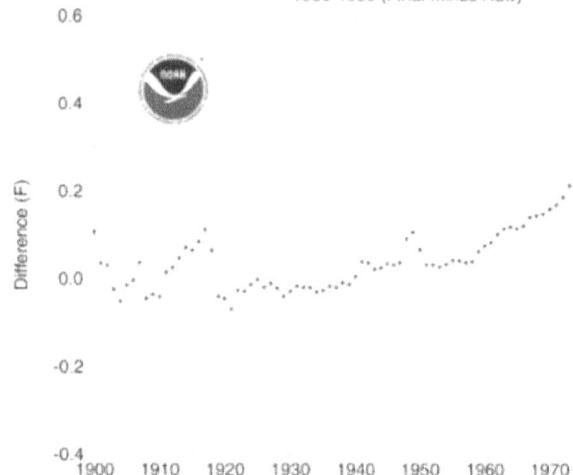

DIFFERENCE BETWEEN RAW AND FINAL USHCN DATA SETS
1900-1999 (Final minus Raw)

Figure 52 - Courtesy of NOAA. The difference between raw and final data sets as computed by the NOAA. Clearly, a warming has been introduced in recent times.

We can clearly see that NOAA's adjustments to raw temperature data have generally been to increase, not decrease, recent temperatures. The net effect of NOAA's adjustments is to increase the rise in temperature since 1900 by 0.5' Fahrenheit.

There are at least three things fundamentally wrong with adjusting site data:

The first is that the recent data is always adjusted upwards to show more warming, and early 20th century temperatures are always adjusted downwards. If two data sets that should match don't, then they adjust the lower data set upwards to match the higher one. All this does is impart more warming into the data.

The second is that the NOAA has never themselves verified the quality of the surface stations. They are clueless to which stations can possibly be giving good data and which ones are giving bad data. The only people who have ever put forth the effort to rate the surface stations are Anthony Watts and those associated with his SurfaceStations.org project.

The third would be that it is complete nonsense to do so. How do you write a mathematical equation to adjust for temperature bias when you have a barbeque pit located underneath the thermometer? Would that equation need to know if you cooked chicken more often than beef? Or if you prefer your steak medium or well done?

Clearly, the only solution is to ensure that all stations are properly sited, and that temperature measurements are taken regularly. Only then can we begin to trust the U.S. temperature data.

If you are wondering what the ClimateGate scientists are saying privately about the matter, we have the answer:

"*Phil Jones,*

We probably need to say more about this. Land warming since 1980 has been twice the ocean warming — and skeptics might claim that this proves that urban warming is real and important.

Comments?"

- Tom Wigley[58]

You're damn right we are noticing. Behind closed doors we see that they are fully aware that they are feeding us nonsense. These are your tax dollars that fund most of this research.

In Figure 53 is the previously mentioned email to James Hansen which was recovered from NASA via a Freedom of Information Act. This is the smoking gun for admitting temperature adjustments.

We see a discussion between Makiko Sato and James Hansen in which they are reviewing the mean temperature data for the years 1934 and 1998. We see that Sato clearly states that at some point he had 1998 warmer than 1934, and we see the progression they followed of slowly decreasing the temperature of 1934 while increasing the temperature of 1998.

So, after we consider all of these facts, do you think the U.S. temperature data should be used to write legislation? I don't.

From: Makiko Sato <makis@███████>
To: James Hansen <jhansen@███████>, Reto Ruedy <rruedy@███████>
Subject: Re: Fwd: FW: <no subject>
Date: Tue, 14 Aug 2007 14:09:34 -0400

I am sure I had 1998 warmer than 1934 at least
once because on my own temperature web page
(which most people never look at), I have

US annual

(Last Modified: 2007-01-12)

and since it was made in January when I updated
all the graphs, I had my US mean table which is
consistent with this until last Monday.

I didn't keep all the data, but some of them are
                    1934   1998
    1999 July   1.459   0.918
    2000 Nov.   1.273   1.151
    2001 Jan.   1.235   1.199   <= These
changes in early years may be due to different analysis
    2006 Jan.   1.235   0.930   <= This
is questionable, I may have kept some data which I was checking.
    2007 Jan.   1.227   1.242   <= This
is only time we had 1998 warmer than 1934, but one web for 7 months.
    2007 Mar.   1.247   1.234   <= Somehow
I recomputed in March, but didn't make changes to the web page.
    2007 Aug.   1.249   1.226   <= Most
recent with corrections, and with July data

I am sorry, I should have kept more data, but I
was not interested in US data after 2001 paper.

Makiko

Figure 53 - Makiko Sato and James Hansen conversing about altering the temperatures for two of the warmest years of the 20th century

Chapter 10

More On Bad Temperature Data

"Anyone who watches BBC News will be left in no doubt that virtually every flood, earthquake, drought or unusual natural occurrence around the world is a direct consequence of global warming...It is very difficult to have a grown-up discussion when a moral crusade, such as the one around climate change, is presented to the public as factual news."

- Frank Furedi, Planet Relief: The Crusade Against Open Debate

In the previous chapter I discussed the poor quality temperature data that is found in the United States and showed that the data is manipulated to give it a warming trend. That is only the beginning of the poor quality of data that climatologists collect about the world. There are numerous other problems with the data collection and processing methods. We will cover some of them in this chapter.

LAND DATA VS. SATELLITE DATA

Let's see how bad data compares to good data.

There are basically two primary ways that we collect current temperature information from the earth:

The first method is based on land and sea temperatures, in which we use thermometers to measure the temperature. The temperatures are gathered from stations and ships all around the world and then the results are compiled into different data sets.

As we saw in the previous chapter, these data sets are then adjusted and manipulated in order to account for missing data and to impose a warming bias into the data. One such data set is the HadCRUT3v data set, which is maintained by the Hadley Centre and the Climatic Research Unit of East Anglia University.

The second method is based on satellite temperature measurements of the lower troposphere. Satellites do not measure temperature directly. They measure radiances in various wavelength bands, which must then be mathematically inverted to obtain indirect inferences of temperature. Satellite data is remarkably accurate and is not be affected by the urban heat problems from which land-based measurements suffer. One such data set is from the University of Alabama in Huntsville (UAH), primarily overseen by John Christy.

We will look at the period of warming that alarmists claim was caused by human produced greenhouse gases in both the land-based and satellite-based data sets.

The satellite data starts in 1979 and runs through 2010. We will look at the "warming period" from 1979 - 1998, comparing the slopes of the land-based HadCRUT3v data to the satellite-based UAH data.

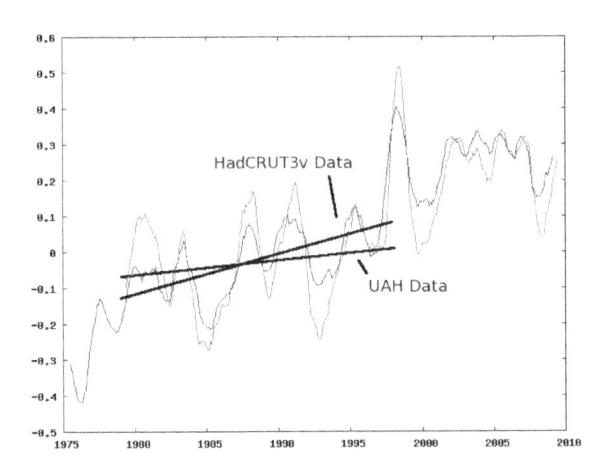

Figure 54 - HadCRUT3v (land) temperature data vs. UAH (satellite) temperature data.

Data Set	Slope (°C/year)	Slope (°C/century)
HadCRUT3v (land)	0.011	1.1
UAH (satellite)	0.004	0.4
Possible Error	0.007	0.7

Table 2 - Slopes of HadCRUT3v (land) data vs. UAH (satellite) data, and the difference bewteen the slopes.

If there are any concerns about cherry picked data, rest assured. Both data sets have been cherry picked for the same years. Unfortunately we can only go back to 1979 because that is when the satellite data starts.

Looking at the numbers, we can see that the HadCRUT3v data has a slope 2.75 times greater than that of the UAH data. This means that the land-based HadCRUT3v data indicates a rate of warming 2.75 times greater than that of the satellite-based UAH data for this period. The alarmist claim of incredible rates of warming disappears in a puff of logic. This clearly allows us to see the influence of their adjusting earlier station records downwards and recent station records upwards to give the appearance of a greater warming trend.

In the bottom row of the table I calculated the possible surface temperature error. To do this I subtracted the UAH data from the HadCRUT3v data. The difference between the two data sets is the amount of warming that has possibly been introduced by all of the problems with station data. For this period of time, we can clearly see that the possible surface error is on the order of 0.7'C per century, which is greater than the amount of warming we have supposedly seen this century.

We can see this trend if we plot the entire set of satellite data too, as shown below. Clearly, The HadCRUT3v data does not represent the real world.

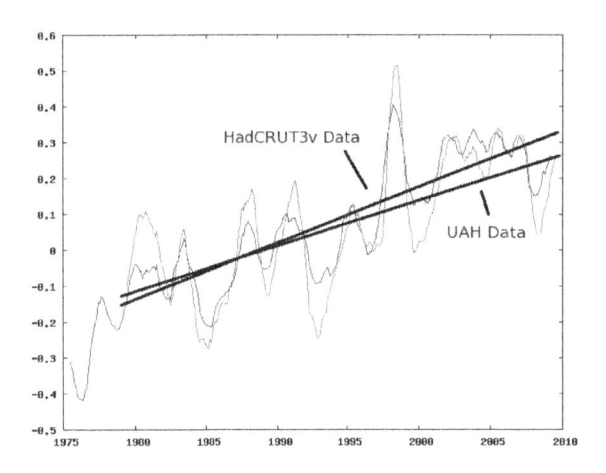

Figure 55 - The complete set of UAH satellite data vs. the HadCRUT3v data

THE BOLIVIA EFFECT

The Bolivia Effect is a term to describe a defect of the temperature record due to absent surface stations. It was originally coined by blogger E. M. Smith[59]. We will describe it here briefly. The following temperature anomaly map was produced for November 2009 using information from the GISTEMP (a land-ocean global mean data set from NASA) data set.

An anomaly map is a map in which temperature anomalies are plotted based on how much warmer or cooler they are than are than a baseline temperature (in this case the average from 1951 – 1980.)

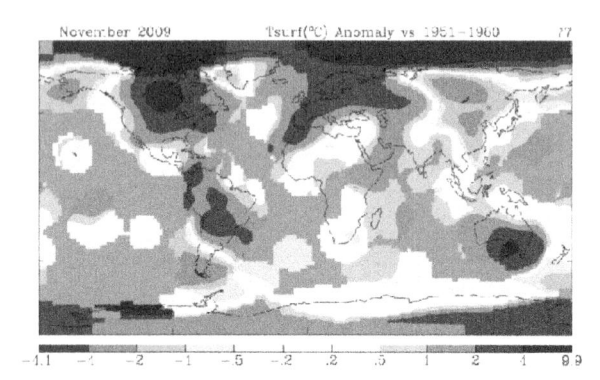

Figure 56 - Temperature anomaly map for November 2009 using GISTEMP data.

All the dark areas indicate warmth. This means that the darker the area, the more warmth that location supposedly experienced. And the lighter the area the cooler the location supposedly experienced. If you look over South America, you will see a rather large dark patch, approximately where Bolivia is located. This is the area we will be focusing on.

Since a very dark spot covers Bolivia, you would be lead to believe that Bolivia was reporting unusually warm temperatures. Unfortunately this is not the case. Bolivia has not had a thermometer that reported data to the Global Historical Climate Network (GHCN) since 1990. So how can it be so hot in Bolivia if there has not been any data from there in the last 20 years?

The answer is unsettling. Just as we saw in the previous chapter, we again see that the data is fabricated. GISTEMP looks for nearby thermometers to get information relevant for Bolivia. The problem is that the nearest thermometer to Bolivia is 1200 kilometers away. That far away from Bolivia includes the beaches of Chile, Peru, and the Amazon jungle.

Regarding Bolivia[60]:

"The Altiplano Region typically has a chilly climate and is considered to have a semi-arid climate. Since it is at a high altitude the thin air retains little heat and the air is typically dry, with cool temperatures and strong cold winds that can sweep over the region.

The geography of Bolivia is unique among the nations of South America. Bolivia is one of two landlocked countries on the continent, and also has the highest average altitude. The main features of Bolivia's geography include the Altiplano, a highland plateau of the Andes, and Lake Titicaca (Lago Titicaca), the largest lake in South America and the highest commercially navigable lake on Earth (which it shares with Peru).

Temperatures and rainfall amounts in mountain areas vary considerably. The Yungas, where the moist northeast trade winds are pushed up by the mountains, is the cloudiest, most humid, and rainiest area, receiving up to 152 cm (60 in) annually. Sheltered valleys and basins throughout the Cordillera Oriental have mild temperatures and moderate rainfall amounts, averaging from 64 cm (25 in) to 76 cm (30 in) annually. Temperatures drop with increasing elevation, however. Snowfall is possible at elevations above 2,000 m (6,562 ft), and the permanent snow line is at 4,600 m (15,092 ft). Areas over 5,500 m (18,045 ft) have a polar climate, with glaciated zones. The Cordillera Occidental is a high desert with cold, windswept peaks."

Smith drives home the point that without thermometers in these different parts of Bolivia it is impossible to accurately represent the varying temperatures present there. You can't take a reading 1200 kilometers away and assume that it accurately represents Bolivia. It is deceitful to represent Bolivia on a map as reporting warmer climate when they are in fact not reporting anything at all.

Sadly, the "Bolivia Effect" is not restricted to one part of South America. It is found throughout the globe.

Looking back at the anomaly map, we see a rather large dark (warm) spot over Canada. In this area, there are no thermometers above the 65' North latitude line in the Yukon and Northwest Territories in the GISTEMP data.

Only one thermometer survives in this area. It is located in Nunavut. The location of this sole thermometer is an area called "The Garden Spot of the Arctic" due to the unusual warmth of the area allowing a variety of plants and animals to survive there that do not survive elsewhere. Worse, the station is located near water. Water serves to buffer the extremes of temperature.

So this one thermometer in an abnormally warm spot of Canada gives data for the entire northern part of the country. How can we possibly accept this as an accurate representation of the temperatures in the area?

Lastly, if we look at the Arctic we see a lot of "warmth" too. Unfortunately, this is all fictional as well. No thermometers exist in the Arctic in the GISTEMP data. The temperatures for this area are created based on nearby stations in warmer areas, and are calculated based on interpolations of other data. Smith muses that in the end, it is all "just made up."

Clearly, the methods currently employed by climate centers to estimate temperatures of the globe are deficient. We are in need of a major overhaul of the methods used to handle missing temperature station data. It is much better to have a blank spot on a map than it is to have unreliable, false data.

Chapter 11

Hide the Decline

"I've just completed Mike's Nature trick of adding in the real temps to each series for the last 20 years (ie from 1981 onwards) and from 1961 for Keith's to hide the decline."

- Phil Jones, University of East Anglia Climatic Research Unit

The head of University of East Anglia's Climatic Research Unit, Phil Jones, wrote the most infamous of the ClimateGate emails:

"Ray Bradley, Michael Mann, Malcolm Hughes, Keith Briffa, and Tim Osborn,

I've just completed Mike's Nature trick of adding in the real temps to each series for the last 20 years (ie from 1981 onwards) and from 1961 for Keith's to hide the decline. Mike's series got the annual land and marine values while the other two got April-Sept for NH [Northern Hemisphere] land North of 20'N. The latter two are real for 1999, while the estimate for 1999 for NH combined is +0.44C with respect to 61-90. The Global estimate for 1999 with data through Oct is +0.35C cf. 0.57 for 1998.

Phil Jones[61]*"*

In this chapter we will briefly describe what Phil Jones claimed "trick" and "hide the decline" meant, then show what it really meant, and then show why this email is so important

First, we will look at Phil Jones' official explanation from November 23, 2009 for his use of the word "trick":

"The word 'trick' was used here colloquially as in a clever thing to do. It is ludicrous to suggest that it refers to anything untoward." - Phil Jones

97

So Phil Jones' position is that the word "trick" did not mean to imply anything deceitful. That is the whole of his attempt to explain this incriminating email. Jones did not address "hide the decline" in this explanation. Alarmist apologists have furthered Jones' explanation by saying that the "hide the decline" was an attempt to deal with the "divergence problem."

THE DIVERGENCE PROBLEM

What is the divergence problem? Before I explain that, I need to give you a very brief introduction to the science of dendrochronology.

Dendrochronology (dendro = tree; chron = time; ology = study of) is the scientific method of dating based on the analysis of patterns of tree-rings. Dendrochronologists study tree rings and attempt to reconstruct past temperature and climate data from them. Since we did not have accurate thermometer records prior to 1850-ish we have to rely on other sources of data for temperature information. These other sources are called proxies.

Proxy records contain a signal that corresponds to climate, but that signal may be weak and embedded in a great deal of background noise. Deciphering that record is often a complex business. Examples of proxy records include ice cores, lake sediments, ocean sediments, and tree rings.

To determine the past temperature, the dendrochronologist cuts a core out of a tree and then looks at the width of the tree rings. They trust that a wider tree ring indicates a greater growing season, which they believe is primarily driven by a higher temperature. What has not been discussed is how they disentangle other factors that influence tree growth, such as amount of sunlight, fertilization, and carbon dioxide. As you may have guessed, there is no known method to disentangle this information. Regardless, they assume that wider tree rings indicate a higher temperature, instead of it possibly meaning a greater exposure to sunlight, greater availability of fertilizer, or an increase in carbon dioxide.

To see if these tree ring temperature reconstructions are accurate they are validated against the past 160 years of thermometer temperature records. See the figure below for an example. The three lines that run the full length of the graphic are the tree ring proxy reconstructions. The line on the right side of the graphic is the thermometer temperature record:

More On Bad Temperature Data

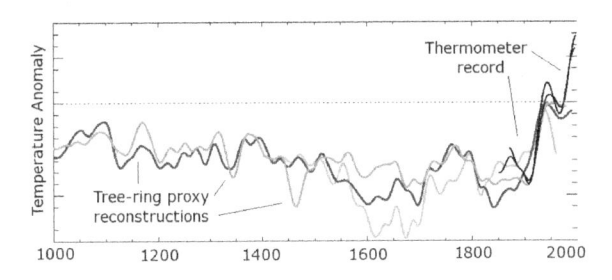

Figure 57 – Tree ring proxy temperature reconstructions plotted with the thermometer temperature data.

You can look at this and judge for yourself whether the tree ring reconstructions match up reasonably well with the thermometer temperature records. They appear to match up quite well from 1900-1950, where the graphs ramp upwards. Around 1960 the tree ring reconstructions greatly diverge from the thermometer record. You can see the light gray line declines while the thermometer record ascends. You can also see that the other two tree ring lines diverge from the temperature record, although they don't dip quite as bad as the other. This non-agreement of the tree ring reconstructions with the thermometer record is the "divergence problem."

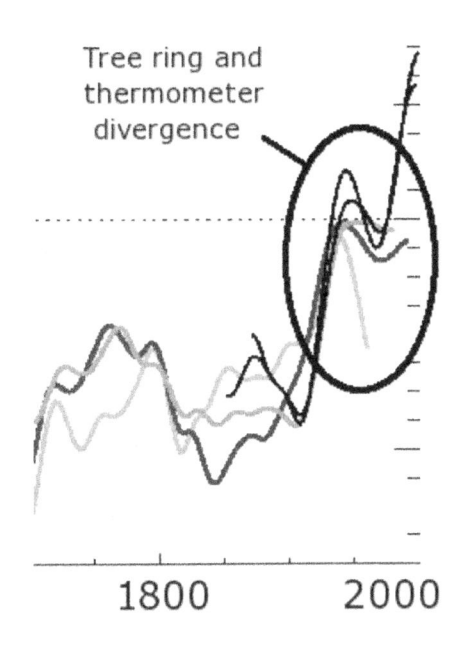

Figure 58 – Close up of the tree ring proxy reconstruction and thermometer divergence.

It is important to see the divergence problem in context. The tree ring reconstructions diverge from the thermometer records primarily during the most recent warming period that we experienced from 1975 – 1998. This means that the trees were unable to resolve the most recent period of warming. If these sets of tree ring proxies are unable to resolve warming in recent times, then we can not trust the data to resolve equally warm temperatures in the past.

This is damaging to the idea of unprecedented catastrophic anthropogenic global warming because these tree ring proxies are not scientifically capable of showing past temperatures that rival those of our most recent warming period. This is the equivalent of taking a black and white photograph to show the beauty of a rainbow.

To date, dendrochronologists have not been able to reconcile this divergence problem of the tree ring record from the thermometer record. This fact is the biggest skeleton in the closet for dendrochronology. As we will see, they are well aware of this problem and yet they ignore it and push their results on the public.

This ClimateGate email on the subject is rather telling:

"Jonathan Overpeck,

Do you know anything about the "divergence problem" in tree rings? Rosanne D'Arrigo talked to the National Research Council yesterday. I didn't get to talk to her afterward, but it looked to me that they have re-drilled a bunch of the high-latitude tree rings that underlie almost all of the high-resolution estimates, and the tree rings are simply missing the post-1970s warming, with reasonably high confidence. She didn't seem too worried, but she apparently has a paper just out in the Journal of Geophysical Research.

It looked to me like she had pretty well killed the "hockey stick" graph in public forum — they go out and look for the most-sensitive trees at the edge of the tree line, flying over lots and lots of trees that are less sensitive but quite nearby, and when things get a little warmer, the most-sensitive trees aren't sensitive any more; and so the trees miss the extreme warming of the recent times, and can't reliably be counted as catching the extreme warmth of the Medieval Warm Period if there was extreme warmth then."

- Richard Alley[62]

We have shown what Phil Jones claimed he meant by using a "trick" to "hide the decline." We have shown that "hide the decline" was meant to deal with the divergence problem, and we have shown that dendrochronology indeed suffers from the divergence problem, and that because of this we can not accept their tree ring proxy reconstructions to accurately give us important temperature information from the past.

Let's look at what Phil Jones actually meant by "hide the decline." This becomes complicated, because his one sentence is so telling and actually covers so much material.

"I've just completed Mike's Nature trick of adding in the real temps to each series for the last 20 years (ie from 1981 onwards) and from 1961 for Keith's to hide the decline"

He is saying that he completed Mike's trick. This means that there are two tricks: Mike's trick, and Phil's trick.

MIKE'S TRICK

First we will look at Mike's trick. Mike in this case is Michael Mann, the inventor of the now-debunked hockey stick, and probably the biggest bully (as revealed in the ClimateGate emails) in climate science. Michael Mann's "trick" has been expertly discussed on the Internet by Steve McIntyre of ClimateAudit[63]. I invite you to read his page for a very thorough and damning explanation. I will walk through a very brief explanation of Mike's trick.

A graphic was needed for the Intergovernmental Panel on Climate Change's (IPCC) Third Assessment Report (TAR). The IPCC reports are supposed to assess scientific information relevant to human-induced climate change, the impacts of human-induced climate change, and options for adaptation and mitigation. These reports are viewed by legislators as the definitive source for global warming information and are used to write laws controlling our production of greenhouse gases.

The people associated with writing this report wanted a graphic that contained several different tree ring reconstructions that illustrated two things: 1) that the recent period of warming was unprecedented, and 2) that temperatures were spiking upwards in recent times. The first graph they assembled was the one you see below.

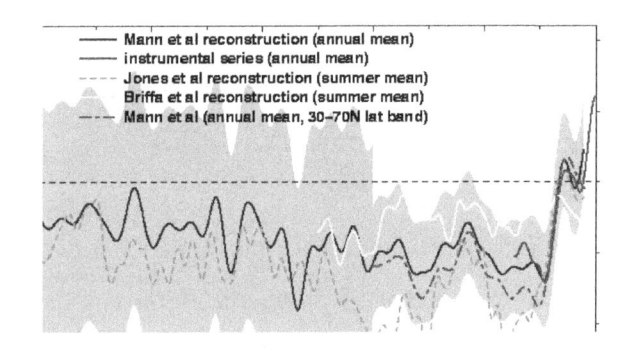

Figure 59 – The original graphic presented for inclusion in the IPCC TAR.

You can see that Keith Briffa's data (in very light gray) did not match up well with the data from Mann and Jones. Briffa's data is the line that rests above the other ones, and if you look you can see it decline in more recent times. Briffa was pressured to alter his graph to show more global warming, as evidenced in this email:

"A proxy diagram of temperature change is a clear favourite for the Policy Makers

summary. But the current diagram with the tree ring only data [i.e. the Keith Briffa reconstruction] somewhat contradicts the multiproxy curve and dilutes the message rather significantly.." – Chris Folland[64]

So we see that Briffa had submitted his data, but his data did not convey the message of catastrophic anthropogenic global warming as much as they would like to. The pressure is now on for Keith Briffa to alter his data to make it look more in line with that of Mann and Jones. That is, to make the past look cooler and the present be warmer. Briffa replies:

"Let me say that I don't mind what you put in the policy makers summary if there is a general consensus. However some general discussion would be valuable ... whether this represents 'TRUTH' however is a difficult problem. I know Mike thinks his series is the 'best' and he might be right - but he may also be too dismissive of other data and possibly over confident in his data (or should I say his use of other's data...)

I know there is pressure to present a nice tidy story as regards 'apparent unprecedented warming in a thousand years or more in the proxy data' but in reality the situation is not quite so simple. We don't have a lot of proxies that come right up to date and those that do (at least a significant number of tree proxies) some unexpected changes in response that do not match the recent warming. I do not think it wise that this issue be ignored in the chapter...

I believe that the recent warmth was probably matched about 1000 years ago. I do not believe that global mean annual temperatures have simply cooled progressively over thousands of years as Mike appears to and I contend that that there is strong evidence for major changes in climate over the Holocene that require explanation and that could represent part of the current or future background variability of our climate..."

– Keith Briffa[65]

Much discussion ensued, and eventually Keith Briffa caved and adjusted his data to be more in line with what the message that the other climate scientists wanted to convey. We then ended up with this graphic being published in the report:

More On Bad Temperature Data

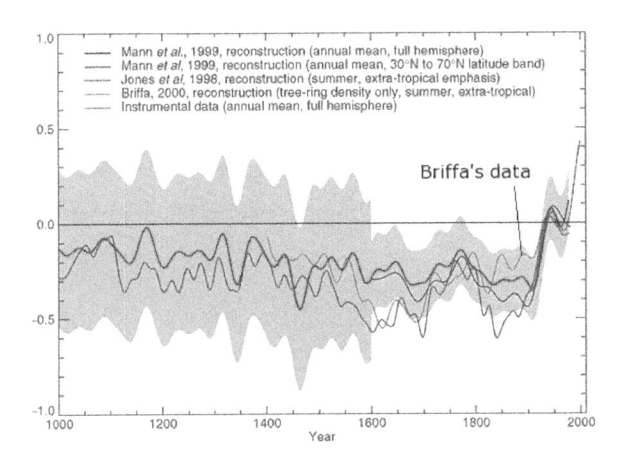

Figure 60 – Tree ring proxy reconstruction data and instrumental thermometer data as presented in the IPCC Third Assessment Report.

If this graph is hard to read, don't worry. This is known as a spaghetti graph because the data looks like limp noodles. If you look rather carefully, you can see that Keith Briffa's data has been moved so that it is more in line with what Mann and Jones had submitted. Also, in this graphic, the decline in Briffa's data has been eliminated. You can clearly see that it trends up with the rest of the data, and then you don't really see it anymore. In other words, the decline has been hidden.

This blow-up of the graphic shows where the decline is hidden:

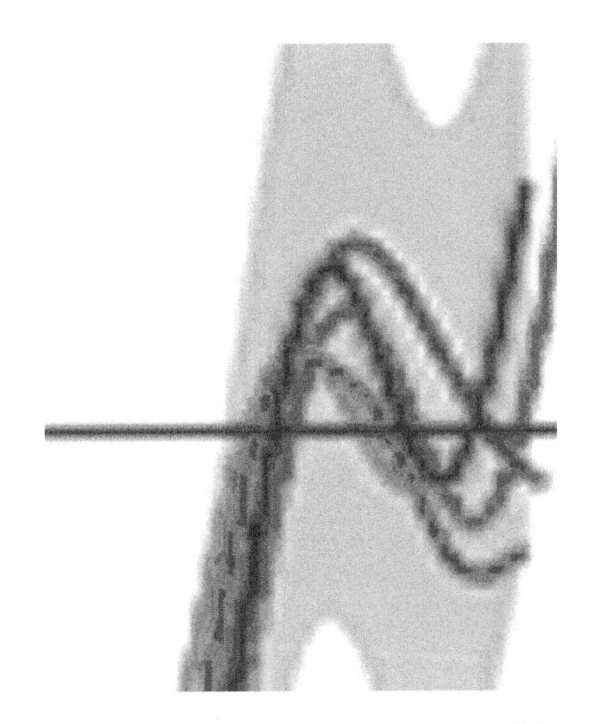

Figure 61 – Close-up of Mann's hiding of Briffa's decline. If you look carefully you can see that 5 sets of data go into the curve, and only four come out of the curve.

You can clearly see that Michael Mann cut off the decline of the graph and hid the truncation of Briffa's data behind the rest of the lines.

If you think this is a misrepresentation of what Mann did, look at the following graphic which displays the before and after of Briffa's data:

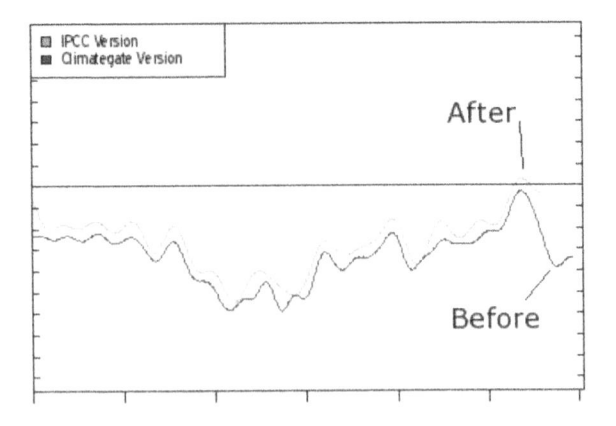

Figure 62 – Before: Keith Briffa's original tree ring reconstruction data with the decline included; After: Keith Briffa's data after it the decline has been chopped and hidden by Michael Mann.

The dark line shows Briffa's original data. The light line shows how Briffa's data was represented in Mann's graphic. You can clearly see that the temperature decline in the reconstruction was truncated.

PHIL'S TRICK

Now, let's look at Phil Jones' trick. Phil Jones was preparing a graphic that would be featured on the cover of an upcoming World Meteorological Organization (WMO) report. This is an image that would be seen by many thousands of people, including legislators that will use this information to make laws that will tax our emissions.

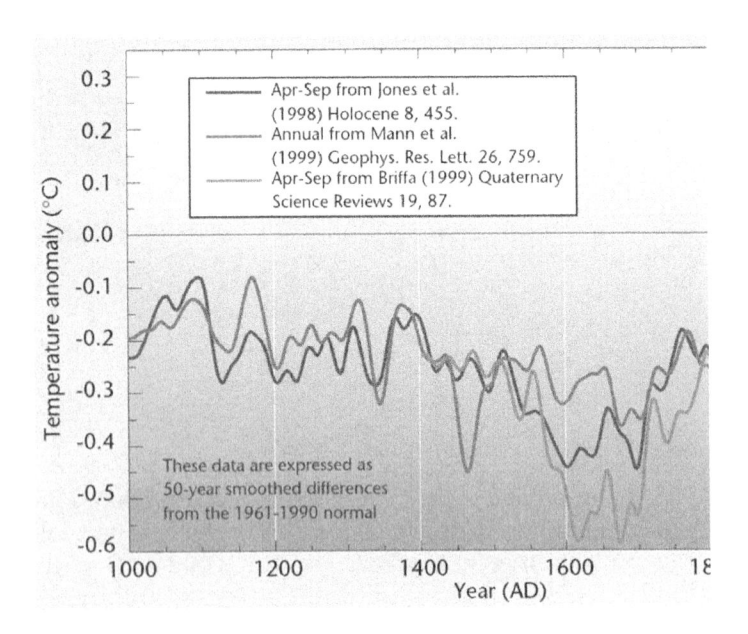

Figure 63 – The image Phil Jones submitted to the World Meteorological Organization for the cover of an upcoming report.

When we look at Jones' image above, we see three lines that go back about 1,000 years. These three lines are tree ring proxy reconstructions. These three lines are the same data from Mann, Jones, and Briffa that were featured in Mann's hidden decline graph that we previously discussed. As the lines move toward the present we see that all of the lines shoot upwards. This conveys the message that these three lines all agree that the recent temperatures skyrocket upwards.

So let's look at the original data he was working with:

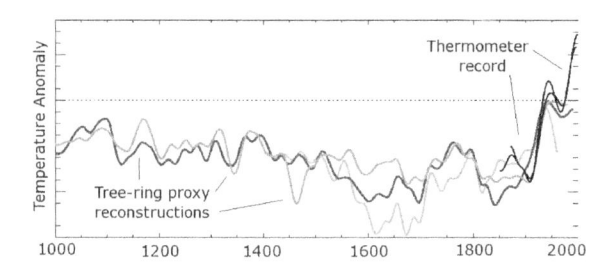

Figure 64 – The data that Phil Jones was working with when he created the World Meteorological Organization report cover

This is the same graphic we saw a few pages back to demonstrate the divergence problem. Note that there are three tree ring proxy reconstructions shown along with the thermometer temperature record. Look at the present time on this graph. If needed, refer to the previous enlargement that demonstrated the divergence of the proxy data from the thermometer data. You clearly can see one line trend down sharply, one line trend flat, and the last line trend down then back up.

Now look at the chart that Phil Jones published for the WMO report. These declines in the temperature data are absent! What happened?

You can plainly see that Phil Jones cut off the declining parts of the tree ring proxy reconstructions and spliced on the thermometer data. Where exactly did he do that? Fortunately we don't have to do any detective work to figure that out. Phil Jones told us:

"I've just completed Mike's Nature trick of adding in the real temps to each series for the last 20 years (ie from 1981 onwards) and from 1961 for Keith's to hide the decline."

You can clearly see that he cut off Briffa's data around 1960 and spliced on the thermometer record, and he cut off his own and Mann's data at around 1980 and spliced on the thermometer record. Notice that none of the tree ring data were indicating that the temperatures were spiking upwards in the same manner that the thermometer instruments were.

Clearly, he "hid the decline" in the tree ring proxy temperature data. Jones' method of cutting off proxy data and splicing thermometer data onto it was never discussed in the World Meteorological Report. Never did he mention that he spliced tree ring data with thermometer data to give the impression that all of the tree ring data suggests recent warming.

Again, the implications of this are that all of their proxy series failed to show the most recent period of warming. Therefore, the proxy data cannot be used to show that the past has not had periods of warming that rival ours. If they cannot detect current warming, then they cannot detect past warming.

THE GROUP COVER-UP

What makes this worse is that Phil Jones sent the "hide the decline" email to Ray Bradley, Michael Mann, Malcolm Hughes, Keith Briffa, and Tim Osborn. This implies that 1) they were all aware of his "trick" and 2) they thought this scientific deceit was acceptable as evidenced by the fact that no one replied to Jones' email letting him know that it was a very deceitful and dishonest thing to do. Apparently presenting fraudulent data was a common enough occurrence that no one was bothered by it, and they felt comfortable emailing their colleagues admitting it.

These co-conspirators were more than happy to go along with Jones' misrepresenting and misleading data. Perhaps they were under too much peer-pressure to go along with presenting a nice and tidy story of catastrophic anthropogenic global warming. Or, perhaps they were equally dishonest.

When challenged on this, Michael Mann posted the following statement to his RealClimate website on Dec 22, 2004:

"No researchers in this field have ever, to our knowledge, 'grafted the thermometer record onto' any reconstruction. It is somewhat disappointing to find this specious claim (which we usually find originating from industry-funded climate disinformation websites) appearing in this forum."[66]

This is a blatant, if not paranoid, lie. Mann was fully aware of this occurring. His colleague Phil Jones had emailed him and specifically explained how he did it. Mann got the email, he saw the report, and he knew it happened. Not only this, but by not stopping Jones from doing so, he essentially endorsed this behavior.

So, there we have it and it could not be any clearer. We have seen how Mann truncated and hid the decline in his graphic, and then we saw how Jones cut off the declining parts of the tree ring reconstructions and grafted the thermometer record onto them. Two of the biggest names in climate science acting deceitfully are caught red-handed. And worse, they are caught by their own

More On Bad Temperature Data

admission. These are your tax dollars in action.

Chapter 12

Bullying, Hiding, And The Money

"The grubby, petty dishonesty disclosed by the [ClimateGate] emails, the apparent willingness to let the world's policies be determined by a political, rather than scientific, agenda, the ambition to be not just the top, but the only view on display - I've seen these things all my adult life in offices in different countries, in marketers, product managers, CEOs and CFOs, and all manner of consultants...It's silly of me, I suppose, to ever have thought science was different, that scientists weren't as human as the rest of us."

- Thomas Fuller

This chapter will highlight certain ClimateGate emails that demonstrate the bullying tactics certain scientists would use in order to prevent dissenting scientists from getting published. It will demonstrate the efforts of the ClimateGate team to hide their data and to delete other data, and it will also briefly talk a little about money.

Perhaps the most difficult part of this book will be selecting only a few ClimateGate emails to show you. There are over a thousand of them, and they are all worth a read. All of the emails are available to read on this book's website:

theclimateconspiracy.com/files/climategate/FOIA/mail

BULLYING

Below we see an email from Phil Jones who has just received word that two astrophysicists, Willie Soon and Sallie Baliunas, published a paper in *Climate Research* which concluded "the 20th century is probably not the warmest nor a uniquely extreme climatic period of the last millennium." Outraged, Jones writes:

"Colleagues,

Tim Osborn has just come across this. Best to ignore it probably, so don't let it spoil your day. I've not looked at it yet. It results from this journal having a number of editors. The responsible one for this is a well-known skeptic in New Zealand. He has let a few papers through by (skeptics) Michaels and Gray in the past. I've had words with Hans von Storch about this, but got nowhere.

Writing this I am becoming more convinced we should do something...

I will be emailing the journal to tell them I'm having nothing more to do with it until they rid themselves of this troublesome editor. A Climatic Research Unit person is on the editorial board, but papers get dealt with by the editor assigned by Hans von Storch."

- Phil Jones[67]

Clearly, Jones is threatening to never publish another paper in the journal again merely because they published a paper that was contrary to Jones' ideological views. Keep in mind, that as Phil admitted, he has not even read the paper yet. He is not concerned if the paper is scientifically valid; he only cares if it supports his views or not.

Michael Mann replies to Phil Jones, planning out a course of damage control. Mann has not read the paper either, and yet he is already so upset by the fact it was published. Being the arrogant and paranoid scientist he is, Mann believes that skeptics have overtaken the journal *Climate Research* in order to get papers published.

Notice at the end, like Jones, Mann suggests blackballing the journal. They are literally trying to damage a journal merely because it published a single paper out of line with their views.

"Phil Jones,

The Soon and Baliunas paper couldn't have cleared a "legitimate" peer review process anywhere. That leaves only one possibility – that the peer-review process at Climate Research has been hijacked by a few skeptics on the editorial board. And it isn't just De Freitas; unfortunately, I think this group also includes a member of my own department... The skeptics appear to have staged a "coup" at Climate Research (it was a mediocre journal to begin with, but now it's a mediocre journal with a definite "purpose").

I told Mike MacCracken that I believed our only choice was to ignore this paper. They've already achieved what they wanted – the claim of a peer-reviewed paper. There is nothing we can do about that now, but the last thing we want to do is bring attention to this paper, which will be ignored by the community on the whole…

It is pretty clear that the skeptics here have staged a bit of a coup, even in the presence of a number of reasonable folks on the editorial board (Whetton, Goodess, …). My guess is that Von Storch is actually with them (frankly, he's an odd individual, and I'm not sure he isn't himself somewhat of a skeptic himself), and with Von Storch on their side, they would have a very forceful personality promoting their new vision.

There have been several papers by Pat Michaels, as well as the Soon and Baliunas paper, that couldn't get published in a reputable journal.

This was the danger of always criticizing the skeptics for not publishing in the "peer-reviewed literature". Obviously, they found a solution to that – take over a journal!

So what do we do about this? I think we have to stop considering Climate Research as a legitimate peer-reviewed journal. Perhaps we should encourage our colleagues in the climate research community to no longer submit to, or cite papers in, this journal. We would also need to consider what we tell or request of our more reasonable colleagues who currently sit on the editorial board…"

- Michael Mann[68]

Phil Jones replies to Mann, stating that he would like to get a paper together which redefines the historical periods of the Medieval Warm Period and the Little Ice Age, and then have all their colleagues sign it. Again, science does not work by a democracy. Science works based on facts. Jones is not concerned by the fact that none of their science supports the idea of an attenuated Medieval Warm Period and Little Ice Age. He is solely interested in pushing his agenda on people. The science is irrelevant to him.

Michael Mann,

Can we not address the misconceptions by finally coming up with definitive dates for the Little Ice Age and Medieval Warm Period and redefining what we think the terms really mean? With all of us and more on the paper, it should carry a lot of weight. In a way we will be setting the agenda for what should be being done over the next few years.

- Phil Jones[69]

Next we see an email from Tom Wigley in which he is upset because another paper skeptical of global warming was published. We see that Wigley believes that only his small group of climate scientists should be allowed to decide what is and what is not published.

In the last paragraph we see that Wigley clearly states that the skeptical scientists have "genuine scientific credentials," but that they should not be allowed to publish in peer-reviewed journals merely because they do not agree with global warming dogma. Only in a religion would skepticism be heresy. Cleary, the peer-review process has been bastardized.

"Colleagues,

Danny Harvey and I refereed a paper by skeptic Pat Michaels and coworkers and said it should be rejected. We questioned the editor (de Freitas again!) and he responded, saying:

"The manuscript was reviewed initially by five referees. ... The other three referees, all reputable atmospheric scientists, agreed it should be published subject to minor revision. Even then I used a sixth person to help me decide. I took his advice and that of the three other referees and sent the manuscript back for revision. It was later accepted for publication. The refereeing process was more rigorous than usual."

On the surface this looks to be above board — although, as referees who advised rejection, it is clear that Danny and I should have been kept in the loop and seen how our criticisms were responded to.

I suspect that de Freitas deliberately chose other referees who are members of the skeptics camp. I also suspect that he has done this on other occasions. How to deal with this is unclear, since there are a number of individuals with genuine scientific credentials who could be used by an unscrupulous editor to ensure that "anti-greenhouse" science can get through the peer review process (Legates, Balling, Lindzen, Baliunas, Soon, and so on). The peer review process is being abused, but proving this would be difficult."

- Tom Wigley[70]

The next day, Wigley plots a way to ensure no skeptical papers are further published in *Climate Research*. He is pushing his colleagues to start a smear campaign against *Climate Research*. Notice that he is not concerned if what he tells the journal is true or not, he only cares that they stop publishing papers that dissent from the idea of anthropogenic global warming.

"Colleagues,

Regarding Climate Research, I do not know the best way to handle the specifics of the editing. Hans von Storch is partly to blame — he encourages the publication of crap science "in order to stimulate debate". One approach is to go direct to the publishers and point out the fact that their journal is perceived as being a medium for disseminating misinformation under the guise of refereed work. I use the word "perceived" here, since whether it is true or not is not what the publishers care about — it is how the journal is seen by the community that counts.

I think we could get a large group of highly credentialed scientists to sign such a letter — 50+ people. Note that I am copying this view only to Mike Hulme and Phil Jones. Mike's idea to get the editorial board members to resign will probably not work — we must get rid of von Storch too, otherwise the holes will eventually fill up with people (skeptics) like Legates, Balling, Lindzen, Michaels, Singer, etc. I have heard that the publishers are not happy with von Storch, so the above approach might remove that hurdle too."

- Tom Wigley[71]

Mann replies. We see him refer to the publication of a single skeptical paper as an "assault." John Costella believes this demonstrates that Michael Mann has an inferiority complex and cannot bear to be criticized by an astrophysicist from Harvard.

"Colleagues,

This might all seem laughable, if it weren't the case that they've gotten the (Bush) White House Office of Science & Technology taking it as a serious matter (fortunately, Dave Halpern is in charge of this project, and he is likely to handle this appropriately, but not without some external pressure).

Here, I tend to concur at least in spirit ... that other approaches may be necessary. I would emphasize that there are indeed, as Tom notes, some unique aspects of this latest assault by the skeptics which are cause for special concern. This latest assault uses a compromised peer-review process as a vehicle for launching a scientific disinformation campaign (often viscious and personal) under the guise of apparently legitimately reviewed science, allowing them to make use of the "Harvard" moniker in the process.

Fortunately, the mainstream media never touched the story (mostly it has appeared in papers owned by Murdoch and his crowd, and dubious fringe on-line outlets). Much like a server which has been compromised as a launching point for computer viruses, I fear that Climate Research has become a hopelessly compromised vehicle in the skeptics' (can we find a better word?) disinformation campaign, and some of the discussion that I've seen (e.g. a potential threat of mass resignation among the legitimate members of the Climate Research editorial board) seems, in my opinion, to have some potential merit.

This should be justified not on the basis of the publication of science we may not like, of course, but based on the evidence (e.g. as provided by Tom and Danny Harvey, and I'm sure there is much more) that a legitimate peer-review process has not been followed by at least one particular editor."

- Michael Mann[72]

In this next email from Mann we see that he is preparing a paper to debunk the Soon and Baliunas paper that previously upset him. I featured this email in the Medieval Warm Period chapter. Remember that Soon and Baliunas provided evidence that the current warm period is not unique, and that it was warmer in the past. This runs contrary to the global warming hypothesis, which maintains that our current warmth is unique in all of history. Mann is now trying to publish a paper to get rid of the warmth of the MWP.

Colleagues,

I think that trying to adopt a timeframe of 2000 years, rather than the usual 1000 years, addresses a good earlier point that Jonathan Overpeck made ... that it would be nice to try to "contain" the putative "Medieval Warm Period", even if we don't yet have data available that far back.

- Michael Mann[73]

We need to realize, again, how much this has bastardized science. Someone published a paper that was not inline with the views of a few scientists, so they respond by launching an attack on the journal, and they go as far as to write an entire paper in an attempt to prove them wrong. They are no longer doing science for the sake of improving human knowledge; they are merely trying to push their global warming dogma on the world. They are performing revenge science.

About a month later, the director of *Climate Research* emailed the complaining scientists:

Colleagues,

In my 20 June 2003 email to you I stated, among other things, that I would ask Climate Research editor Chris de Freitas to present to me copies of the reviewers' evaluations for the two Soon and coworker papers. I have received and studied the material requested. Conclusions:

1) The reviewers consulted (four for each manuscript) by the editor presented detailed, critical and helpful evaluations.

2) The editor properly analyzed the evaluations and requested appropriate revisions.

3) The authors revised their manuscripts accordingly.

Summary: Chris de Freitas has done a good and correct job as editor.

- Otto Kinne, Climate Research[74]

The ClimateGate scientists, defeated, again threaten to blackball the journal:

"It seems to me that this "Kinne" character's words are disingenuous, and probably supports what de Freitas is trying to do. It seems clear we have to go above him. I think that the community should, as Mike Hulme has previously suggested in this eventuality, terminate its involvement with this journal at all levels — reviewing, editing, and submitting, and leave it to wither way into oblivion and disrepute." – Michael Mann[75]

"I agree that Otto Kinne seems like he could be a Chris de Freitas clone. However, what would be our legal position if we were to openly and extensively tell people to avoid the journal?" – Tom Wigley[76]

"Based on Otto Kinne's editorial, I see little hope for more enlightened editorial decision-making at Climate Research. Tom, Richard Smith and I will eventually publish a rebuttal to the Douglass and coworkers paper. We'll publish this rebuttal in the Journal of Geophysical Research — not in Climate Research." – Ben Santer[77]

Moving on, we now find an email where a reporter from The Sunday Telegraph writes to Michael Mann and Phil Jones, asking them to clarify a few points in their most recent paper.

"Michael Mann and Phil Jones,

I'm putting together a piece on global warming, and I'll be making reference to your paper in Geophysical Research Letters with Prof Jones on "Global surface temperatures over the past two millennia".

When the paper came out, some critics argued that the paper actually showed that there have been three periods in the last 2000 years which were warmer than today (one just prior to AD 700, one just after, and one just prior to AD 1000). They also claimed that the paper could only conclude that current temperatures were warmer if one compared the proxy data with other data sets. (For an example of these arguments, see: link to paper)

I'd be very interested to include your rebuttals to these arguments in the piece I'm doing. I must admit to being confused by why proxy data should be compared to instrumental data for the last part of the data-set. Shouldn't the comparison be a consistent one throughout?

With many thanks for your patience with this"

- Robert Matthews[78]

This seems innocent enough. Matthews just wants to fully understand something before he writes an article on it. Mann responds harshly. He lets the reporter know that he has forwarded his email to his ClimateGate colleagues, letting them know to not communicate with Matthews. Effectively, Mann has alienated Matthews from any further communication with any of his group. It is truly bizarre that Mann feels the need to bully and intimidate anyone who questions his work. What a weak, little man.

"Robert Matthews,

Unfortunately Phil Jones is traveling and will probably be unable to offer a separate reply. Since your comments involve work that is his as well, I have therefore taken the liberty of copying your inquiry and this reply to several of his British colleagues.

The comparisons made in our paper are well explained therein, and your statements belie the clearly-stated qualifications in our conclusions with regard to separate analyses of the Northern Hemisphere, Southern Hemisphere, and globe.

An objective reading of our manuscript would readily reveal that the comments you refer to are scurrilous. These comments have not been made by scientists in the peer-reviewed literature, but rather, on a website that, according to published accounts, is run by individuals sponsored by ExxonMobil

Bullying, Hiding, and The Money

Corporation, hardly an objective source of information.

Owing to pressures on my time, I will not be able to respond to any further inquiries from you. Given your extremely poor past record of reporting on climate change issues, however, I will leave you with some final words. Professional journalists I am used to dealing with do not rely upon un-peer-reviewed claims off internet sites for their sources of information. They rely instead on peer-reviewed scientific research, and mainstream, rather than fringe, scientific opinion."

- Michael Mann[79]

Below, Mann has received an email from Steve Mackwell, the Editor in Chief of *Geophysical Research Letters*. He is responding to an issue that Mann brought up in which he was upset because he was unable to review a paper that was critical of his work before it was published.

We can see that Mann feels that no one should be allowed to be critical of his work without him first approving it. Later in the exchange we can see that he truly believes that anyone who is critical of his work is a "contrarian." We will see him tell his colleagues that they should stop publishing in *Geophysical Research Letters* solely because the journal has published papers that are not in line with Michael Mann's views.

"Michael Mann,

In your recent email to Chris Reason, you laid out your concerns that I presume were the reason for your phone call to me last week. I have reviewed the manuscript by McIntyre, as well as the reviews. The editor in this case was Prof. James Saiers. He did note initially that the manuscript did challenge published work, and so felt the need for an extensive and thorough review. For that reason, he requested reviews from three knowledgeable scientists. All three reviews recommended publication.

While I do agree that this manuscript does challenge (somewhat aggressively) some of your past work, I do not feel that it takes a particularly harsh tone. On the other hand, I can understand your reaction. As this manuscript was not written as a Comment, but rather as a full-up scientific manuscript, you would not in general be asked to look it over. And I am satisfied by the credentials of the reviewers. Thus, I do not feel that we have sufficient reason to interfere in the timely publication of this work."

- Steve Mackwell, Editor in Chief of Geophysical Research Letters[80]

Upon receipt of this email, Mann sends this message to his colleagues.

"*Colleagues,*

Just a heads-up. Apparently, the contrarians now have an "in" with Geophysical Research Letters. This guy Saiers has a prior connection with the University of Virginia Department of Environmental Sciences that causes me some unease.

I think we now know how the various Douglass and coworkers papers with Michaels and Singer, the Soon and coworkers paper, and now this one have gotten published in Geophysical Research Letters."

- Michael Mann[81]

Tom Wigley replies, stating that the journal has gone down hill merely because they published papers that do not go along with their idea that climate change is occurring in an unprecedented manner. He clearly states that he believes Saiers is in the "greenhouse skeptic camp."

Keep in mind there is zero evidence that human produced carbon dioxide is contributing to the accelerated warming of our planet. It is accepted solely on faith and computer models. Despite this, these scientists are essentially participating in a witch-hunt to "oust" people who do not accept their shoddy science and piss-poor statistics.

Colleagues,

This is truly awful. Geophysical Research Letters has gone downhill rapidly in recent years. I think the decline began before Saiers. I have had some unhelpful dealings with him recently with regard to a paper Sarah Raper and I have on glaciers – it was well received by the referees, and so is in the publication pipeline. However, I got the impression that Saiers was trying to keep it from being published.

Proving bad behavior here is very difficult. If you think that Saiers is in the greenhouse skeptics' camp, then, if we can find documentary evidence of this, we could go through official American Geophysical Union channels to get him ousted. Even this would be difficult.

- Tom Wigley[82]

Mann continues adding fuel to the fire that *Geophysical Research Letters* has been "taken over" by skeptics because they've published four papers that do not agree with the ClimateGate scientists. Read how they are conspiring to use all of their connections to get the editor fired so that no skeptics can get published. By preventing skeptics from being published this allows the alarmists to make the claim that "all of the peer-reviewed literature supports our ideas."

Colleagues,

Yeah, basically this is just a heads-up to people that something might be up here. What a shame that would be. It's one thing to lose Climate Research. We can't afford to lose Geophysical Research Letters. I think it would be useful if people begin to record their experiences with both Saiers and potentially Mackwell (I don't know him – he would seem to be complicit with what is going on here).

If there is a clear body of evidence that something is amiss, it could be taken through the proper channels. I don't think that the entire American Geophysical Union hierarchy has yet been compromised!

I'm not sure that Geophysical Research Letters can be seen as an honest broker in these debates any more, and it is probably best to do an "end run" around Geophysical Research Letters now where possible. They have published far too many deeply flawed contrarian papers in the past year or so. There is no possible excuse for them publishing all three Douglass papers and the Soon and coworkers paper. These were all pure crap.

There appears to be a more fundamental problem with Geophysical Research Letters now, unfortunately...

- Michael Mann[83]

HIDING

Now we will look at certain ClimateGate emails that deal with hiding data, hiding truths, or other deceitful behavior.

Below we see Phil Jones emailing Michael Mann, describing to him the nature of Freedom of Information Act requests. Jones describes multiple ways that he can hide from the requests in order to avoid having to send his raw data to Steve McIntyre, the Canadian largely responsible for debunking Mann's original hockey stick. Jones goes as far as to threaten to delete the data should he ever be forced to send it. What are you hiding, Jones?

"Michael Mann,

Just sent loads of station data to Scott. Make sure he documents everything better this time! And don't leave stuff lying around on ftp sites - you never know who is trawling them. The two MMs have been after the CRU station data for years.

If they ever hear there is a Freedom of Information Act now in the UK, I think I'll delete the file rather than send it to anyone. Does your similar act in the US force you to respond to enquiries within 20 days? - our does! The UK works on precedents, so the first request will test it.

We also have a data protection act, which I will hide behind. Tom Wigley has sent me a worried email when he heard about it - thought people could ask him for his model code. He has retired officially from the University of East Anglia so he can hide behind that. Intellectual Property Rights should be relevant here, but I can see me getting into an argument with someone at University of East Anglia who'll say we must adhere to it!"

- Phil Jones[84]

Later, Phil Jones sends an email to Michael Mann and friends after hearing that Mann will be forced to release some of his data to other scientists.

"Michael Mann, Ray Bradley, Malcolm Hughes,

The skeptics seem to be building up a head of steam here! I'm getting hassled by a couple of people to release the Climatic Research Unit temperature data. Don't any of you three tell anybody that the United Kingdom has a Freedom of Information Act!

Leave it to you to delete as appropriate!"

- Phil Jones[85]

Below we see Phil Jones emailing Michael Mann, asking him to delete all emails about the latest IPCC report. In context, Jones probably had received a Freedom of Information Act request for those emails and didn't want them getting out. This is illegal.

"Michael Mann,

Can you delete any emails you may have had with Keith Briffa regarding the latest Intergovernmental Panel on Climate Change report? Keith Briffa will do likewise. He's not in at the moment - minor family crisis.

Can you also email Eugene Wahl and get him to do the same? I don't have his new email address.

We will be getting Caspar Ammann to do likewise."

- Phil Jones[86]

Mann responds, going along with the conspiracy:

"Phil Jones,

I'll contact Eugene Wahl about this as soon as possible. His new email is: generwahl@yahoo.com."

- Michael Mann[87]

Next we see Phil Jones emailing Ben Santer, discussing some of the Freedom of Information Act requests that he ignored. Canadian Steve McIntyre sent multiple requests asking for data. They were either ignored or he was sent on a wild goose chase. We see Jones admitting to convincing his colleagues to ignore the FOIA requests and further to admitting to deleting emails that were supposed to be obtainable by such a request.

"Ben Santer,

When the FOI requests began here, the FOI person said we had to abide by the requests. It took a couple of half hour sessions - one at a screen, to convince them otherwise.

The inadvertent email I sent last month has led to a Data Protection Act request sent by a certain Canadian, saying that the email maligned his scientific credibility with his peers!

If he pays 10 pounds (which he hasn't yet) I am supposed to go through my emails and he can get anything I've written about him. About 2 months ago I deleted loads of emails, so have very little - if anything at all."

- Phil Jones[88]

It is a normal and expected thing in publicly funded science to share your data so that other scientists can attempt to reproduce and validate results. For some reason, most likely because he had something to hide, Phil Jones constantly refused to give out his data. Upon being told that he may be required to give out some, he immediately emailed Michael Mann asking for advice and how to avoid giving any out. Notice at the end he is even looking at getting advice from his legal department. Again, what was he hiding?

"Michael Mann,

This is for YOURS EYES ONLY. Delete after reading - please ! I'm trying to redress the balance. One reply from Christian Pfister said you should make all of the data available! Pot calling the kettle black - Christian doesn't make his methods available. I replied to the wrong Christian message so you don't get to see what he said. Probably best. Told Steve separately and to get more advice from a few others as well as Kluwer and legal.

PLEASE DELETE - just for you, not even Ray Bradley and Malcolm Hughes."

- Phil Jones[89]

Below we see Michael Mann emailing Phil Jones. They are discussing hiding data from fellow scientist Steve McIntyre. McIntyre is interested in getting the data from their earlier works and trying to reproduce their results. Knowing that their earlier results are dubious, they conspire to hide the data the best they can. Mann suggests not sending anything, he implies that it will be nearly impossible for McIntyre to get the same results, and lastly he admits if McIntyre had their data it would be to their peril. That doesn't sound very innocent. Again, what makes science great is that other scientists are supposed to be able to reproduce your results. Mann's and Jones' hiding of the data is anti-science.

"Phil Jones,

Personally, I wouldn't send him anything. I have no idea what he's up to, but you can be sure it falls into the "no good" category.

There are a few series from our '03 paper that he won't have--these include the latest Jacoby and D'Arrigo, which I digitized from their publication (they haven't made it publicly available) and the extended western North American series, which they wouldn't be able to reproduce without following exactly the procedure described in our '99 GRL paper to remove the estimated non-climatic component.

I would not give them *anything*. I would not respond or even acknowledge receipt of their emails. There is no reason to give them any data, in my opinion, and I think we do so at our own peril!"

- Michael Mann[90]

Bullying, Hiding, and The Money

Below is an email from 2004 that Mann sent to a couple of his colleagues. In this email he says that he cleaned up some of his computer programs from a paper he wrote in 2003 and is sending it to them in case they want to test it.

We will see at least four things in this email. The first is that it took Mann a whole year after publishing his paper to have his colleagues actually test out his methods. Why did no one check his work prior to publication?

Second, we see that since he stated that he cleaned up the computer program, we can deduce this is not the actual program he published. His published version was not cleaned up, and therefore likely contained errors (as we will see.)

Third, we see paranoid Mann worried that someone who isn't worshiping carbon dioxide may get their hands on the data and discover that its flaws.

Lastly, we see that Mann himself finds an error in his program. This means his published version contained an error. The error served to amplify his results by a factor of 1.29. This sounds alarming. Notice in the email he fails to admit this as an actual error. What an arrogant, little man.

Phil Jones, Tom Crowley, and Gabi Hegerl,

I've attached a cleaned-up and documented version of the computer programs that I wrote for doing the Mann and Jones (2003) calculations. I did this knowing that Phil and I are likely to have to respond to more crap criticisms from the idiots in the near future, so it is best to clean up the programs and provide them to some of my close colleagues in case they want to test it, etc. Please feel free to use these programs for your own internal purposes, but don't pass them along where they may get into the hands of the wrong people.

In the process of trying to clean the programs up, I realized I had something a bit odd, not necessarily wrong, but it makes a small difference. ... It looks like I had two similarly-named data sets floating around in the programs, and used perhaps the less preferable one ....

This may explain part of what perplexed Gabi when she was comparing my results with the real temperatures. I've attached the version of the analysis where the correct data is used instead, as well as the computer programs, which you're welcome to try to use yourself and play around with. Basically, this increases everything everywhere by the factor 1.29. Perhaps this is more in line with what Gabi was estimating (Gabi?).

Anyway, it doesn't make a major difference, but you might want to take this into account in any further use of the Mann and Jones data...

- Michael Mann[91]

Below Phil Jones is replying to an email from Neville Nicholls, of the Bureau of Meteorology Research Centre in Melbourne, Australia. She has asked Jones if he was expecting to get a call from Congress regarding an ongoing investigation. Jones replies that he hopes he doesn't. He is trying to hide the fact that he has been receiving grant money from the U.S. Department of Energy. He hopes that they don't realize they are giving him grant money and that he should be questioned.

Neville Nicholls,

I hope I don't get a call from Congress! I'm hoping that no-one there realizes I have a United States Department of Energy grant, and have had this (with Tom Wigley) for the last 25 years.

- Phil Jones[92]

-

THE MONEY

Now we will look at certain ClimateGate emails that deal with money. This email below is difficult to put into context because we cannot look at their financial records, but it appears that most likely that they ran out of money working on one project without finishing the work. Mick Kelly is telling Nguyen Huu Ninh that they need to prepare a fake set of expenses to make it look like they have money left over so that the NOAA will send them more money, without the NOAA knowing they blew all the money. In the last paragraph we see that they are basically admitting to laundering the money through "Simon's institute."

"Nguyen Huu Ninh,

The NOAA wants to give us more money for the El Nino work with IGCN.

How much do we have left from the last budget? I reckon most has been spent but we need to show some left to cover the costs of the trip Roger didn't make and also the fees / equipment / computer money we haven't spent, otherwise the NOAA will be suspicious.

Politically this money may have to go through Simon's institute but there overhead rate is high so maybe not!"

- Mick Kelly[93]

In the email below we see Russian scientist Tatiana Dedkova asking ClimateGate scientist Keith Briffa to transfer funds to personal accounts and to keep the funds less than 10,000 dollars in order to avoid big taxes. This is called tax evasion. It is rather ironic that they evade taxes, but their research is used to persuade legislators to put taxes on everything we do via cap-and-trade. It's worth nothing that since we do not have Keith's reply, we cannot know if he participated in this tax evasion.

"Keith Briffa,

Also, it is important for us if you can transfer the ADVANCE money on the personal accounts which we gave you earlier and the sum for one occasion transfer (for example, during one day) will not be more than 10,000 USD. Only in this case we can avoid big taxes and use money for our work as much as possible."

- Tatiana M. Dedkova[94]

In the email below we see that Mick Kelly had a meeting with Shell International, and that they are wondering if their new building will be part of their payment. Buildings certainly aren't inexpensive. This one is ironic as well because global warming alarmists always accuse skeptics of being in the pocket of Big Oil, and here we see that Big Oil is buying them a building.

"Michael Hulme, Tim O'Riordan,

Had a very good meeting with Shell yesterday. Only a minor part of the agenda, but I expect they will accept an invitation to act as a strategic partner and will contribute to a studentship fund though under certain conditions. I now have to wait for the top-level soundings at their end after the meeting to result in a response. We, however, have to discuss ASAP what a strategic partnership means, what a studentship fund is, etc, etc. By email? In person?

I hear that Shell's name came up at the TC meeting. I'm CC-ing this to Tim who I think was involved in that discussion so all concerned know not to make an independent approach at this stage without consulting me! I'm talking to Shell International's climate change team but this approach will do equally for the new foundation as it's only one step or so off Shell's equivalent of a board level. I do know a little about the Foundation and what kind of projects they are looking for. It could be relevant for the new building, incidentally, though opinions are mixed as to whether it's within the remit [payment]."

- Mick Kelly[95]

Below we see David Parker, from U.K. Meteorological Office writing to Neil Plummer, from the National Climate Centre of the Bureau of Meteorology, Melbourne, Australia. He is discussing changing the baseline of temperatures in a chapter of the IPCC report so that the graphic would give a greater impression of global warming. In other words, they are manipulating the data in order to show more "global warming."

"Neil Plummer,

There is a preference in the atmospheric observations chapter of the Intergovernmental Panel on Climate Change Fourth Assessment Report to stay with the 1961–1990 baseline. This is partly because a change of baseline confuses users, e.g. anomalies will seem less positive than before if we change to a newer baseline, so the impression of global warming will be muted."

- David Parker[96]

MISCELLANEOUS

Lastly we will look at a few miscellaneous ClimateGate emails. Below we see Adam Markham, from the World Wildlife Foundation, writing to climate scientist Michael Hulme about a paper he wrote regarding climate change in Australia. We see that the agenda-driven activists at WWF are telling climate scientists to make portions of their papers more alarming, so that they can really scare people about climate change.

"Michael Hulme,

I'm sure you will get some comments direct from Mike Rae in WWF Australia, but I wanted to pass on the gist of what they've said to me so far.

They are worried that this may present a slightly more conservative approach to the risks than they are hearing from CSIRO. In particular, they would like to see the section on variability and extreme events beefed up if possible. They regard an increased likelihood of even 50% of drought or extreme weather as a significant risk. Drought is also a particularly

Bullying, Hiding, and The Money

important issue for Australia, as are tropical storms.

I guess the bottom line is that if they are going to go with a big public splash on this they need something that will get good support from CSIRO scientists (who will certainly be asked to comment by the press)."

- *Adam Markham [World Wildlife Foundation]*

Next we see an email from Gavin Schmidt of NASA. He created an Internet website named RealClimate.org. RealClimate is a highly moderated website that solely promotes the alarmist view of climate science. This was to become the propaganda site of the ClimateGate scientists. Basically, the website gives them the ability to do "damage control" for any scientific article that undermines the idea of global warming.

As an example, recently a paper was published by Susan Solomon that identified water vapor as a major driver of temperature change. This undermined the idea of carbon dioxide driving temperature change, so immediately Gavin Schmidt put up a post on RealClimate attempting to lessen the blow the paper gave to the global warming hypothesis. He is truly grasping at straws.

Below is his email announcing the creation of RealClimate.

"Dozens of Colleagues,

No doubt some of you share our frustration with the current state of media reporting on the climate change issue. Far too often we see agenda-driven "commentary" on the Internet and in the opinion columns of newspapers crowding out careful analysis. Many of us work hard on educating the public and journalists through lectures, interviews and letters to the editor, but this is often a thankless task.

In order to be a little bit more pro-active, a group of us (see below) have recently got together to build a new 'climate blog' website: RealClimate.org which will be launched over the next few days.

The idea is that we working climate scientists should have a place where we can mount a rapid response to supposedly 'bombshell' papers that are doing the rounds and give more context to climate related stories or events."

- *Gavin Schmidt*[97]

In 2005 a subcommittee of the U.S. House of Representatives was investigating Michael Mann. Upon finding this out, he reaches out to his colleagues asking for legal advice. If he was innocent, why did he need a legal team?

"Colleagues,

This was predicted--they're of course trying to make things impossible for me. I need immediate help regarding recourse for free legal advice, etc."

- Michael Mann[98]

Tom Wigley replies, cautioning Mann to not use any paleoclimate work outside of the ClimateGate scientists work, because it does not remotely support Mann's now debunked work. Clearly, we see that these scientists do not agree that the most recent warm period was unprecedented.

"Michael Mann,

A word of warning. I would be careful about using other, independent paleoclimatology ... work as supporting your work. I am attaching my version of a comparison of the bulk of these other results. Although these all show the "hockey stick" shape, the differences between them prior to 1850 make me very nervous. If I were on the greenhouse deniers' side, I would be inclined to focus on the wide range of paleoclimatology results and the differences between them as an argument for dismissing them all."

- Tom Wigley[99]

Below we see Phil Jones emailing John Christy, essentially wishing death upon the planet by hoping that catastrophic global warming claims come true. For the sole purpose of being proven right. Clearly this allows us to see into Jones' mind and understand why he was so willing to "hide the decline", and so willing to go along with fraudsters such as Michael Mann, and why he was so willing to hide the fact that he knew global temperatures have been declining since 1998 (See the "Recent Cooling" chapter.)

"John Christy,

As you know, I'm not political. If anything, I would like to see the climate change happen, so the science could be proved right, regardless of the consequences. This isn't being political, it is being selfish."

- Phil Jones[100]

Below we see Phil Jones emailing Michael Mann. First, we see Jones state that he is reviewing a paper that is attacking his own data from the CRU. Of course he is going to say the paper is rubbish and attempt to have it rejected. Next, we see Jones essentially asking Mann in confidence to subvert the peer-review system.

"Michael Mann,

Just agreed to review a paper for Geophysical Research Letters - it is absolute rubbish. It is having a go at the CRU temperature data - not the latest version, but the one you used in 1998. We added lots of data in for the region this person says has urban warming! Such an easy review to do.

Can I ask you something in confidence — don't email around, especially not to Keith and Tim here. Have you reviewed any papers recently for Science that say that the paper by Mann, Bradley, and Hughes in 1998 and the paper by Mann and Jones in 2003 have underestimated variability in the thousand-year record — from models or from some slowly varying temperature proxy data? Just a yes or no will do. Tim is reviewing them — I want to make sure he takes my comments on board, but he wants to be squeaky clean with discussing them with others. So forget this email when you reply."

- Phil Jones[101]

Sorry Phil, we will never forget your emails.

References

[1] http://www.junkscience.com/Greenhouse/

[2] Solomon, S. 2009. Contributions of Stratospheric Water Vapor to Decadal Changes in the Rate of Global Warming. Science DOI: 10.1126/science.1182488

[3] http://assassinationscience.com/climategate/

[4] http://tigger.uic.edu/~pdoran/012009_Doran_final.pdf

[5] http://www.petitionproject.org/

[6] http://www.theclimateconspiracy.com/files/climategate/FOIA/mail/1062592331.txt

[7] http://strata-sphere.com/blog/index.php/archives/11861

[8] http://www.theclimateconspiracy.com/files/climategate/FOIA/mail/0938031546.txt

[9] http://en.wikipedia.org/wiki/Rajendra_Pachauri#Background

[10] http://www.theclimateconspiracy.com/files/climategate/FOIA/mail/1062592331.txt

[11] http://www.nipccreport.org/

[12] http://www.globalwarminghoax.com/print.php?news.123

[13] http://www.financialpost.com/opinion/story.html?id=2197207&p=1

[14] http://epw.senate.gov/public/index.cfm?FuseAction=Hearings.Testimony&Hearing_ID=bfe4d91d-802a-23ad-4306-b4121bf7eced&Witness_ID=6b57de26-7884-47a3-83a9-5f3a85e8a07e

[15] http://www.theclimateconspiracy.com/files/climategate/FOIA/mail/1054757526.txt

[16] http://www.theclimateconspiracy.com/files/climategate/FOIA/mail/1067194064.txt

[17] http://www.theclimateconspiracy.com/files/climategate/FOIA/mail/1067194064.txt

[18] http://www.theclimateconspiracy.com/files/climategate/FOIA/mail/1067532918.txt

[19] http://www.theclimateconspiracy.com/files/climategate/FOIA/mail/0938018124.txt

[20] Mann et al. "Global Signatures and Dynamical Origins of the Little Ice Age and Medieval Climate Anomaly." 2009. Science 326 (5957): 1256-1260.

[21] Soon, W., & Baliunas, S. (2003). Proxy Climatic and Environmental Changes of the Past 1000 Years. Climate Research, 23, 89-110.

[22] http://www.co2science.org/

[23] Tyson, P.D., Karlen, W., Holmgren, K. and Heiss, G.A. 2000. The Little Ice Age and medieval warming in South Africa. South African Journal of Science 96: 121-126.

[24] Goni, M.A., Woodworth, M.P., Aceves, H.L., Thunell, R.C., Tappa, E., Black, D., Muller-Karger, F., Astor, Y. and Varela, R. 2004. Generation, transport, and preservation of the alkenone-based U37K' sea surface temperature index in the water column and sediments of the Cariaco Basin (Venezuela). Global Biogeochemical Cycles **18**: 10.1029/2003GB002132.

[25] von Gunten, L., Grosjean, M., Rein, B., Urrutia, R. and Appleby, P. 2009. A quantitative high-resolution summer temperature reconstruction based on sedimentary pigments from Laguna Aculeo, central Chile, back to AD 850. The Holocene 19: 873-881.

[26] http://en.wikipedia.org/wiki/Vikings
[27] http://ruby.fgcu.edu/courses/twimberley/EnviroPhilo/Glacial.pdf

[28] http://alpen.sac-cas.ch/html_d/archiv/2004/200406/ad_2004_06_12.pdf

[29] http://translate.google.com/translate?hl=de&sl=de&tl=en&u=http%3A%2F%2Fwww.science-skeptical.de%2Fblog%2Fbeispiellose-erwarmung-oder-beispiellose-datenmanipulation%2F001195%2F

[30] Hubert H. Lamb: Climate and cultural history, Reinbeck 1989

[31] Wilfried Weber, The development of the northern limit of viticulture in Europe.

[32] http://translate.google.com/translate?hl=de&sl=de&tl=en&u=http%3A%2F%2Fwww.science-skeptical.de%2Fblog%2Fbeispiellose-erwarmung-oder-beispiellose-datenmanipulation%2F001195%2F

[33] http://en.wikipedia.org/wiki/Sun

[34] http://en.wikipedia.org/wiki/Maunder_Minimum

[35] http://en.wikipedia.org/wiki/Maunder_Minimum

[36] http://en.wikipedia.org/wiki/River_Thames_frost_fairs

[37] Usokin, I.G., Solanki, S., Schussler, M., Mursula, K., Alanko, K. 2003. Millennium-Scale Sunspot Number Reconstruction: Evidence for an Unusually Active Sun since the 1940s. Physical Review Letters: 91.211101.

References

[38] http://www.theclimateconspiracy.com/files/images/2009/12/pachauri_letter.pdf

[39] http://www.theclimateconspiracy.com/files/climategate/FOIA/mail/1255352257.txt

[40] http://www.theclimateconspiracy.com/files/climategate/FOIA/mail/1120593115.txt

[41] http://en.wikipedia.org/wiki/Antarctica

[42] http://www.news.com.au/antarctic-ice-is-growing-not-melting-away/story-0-1225700043191

[43] http://earthobservatory.nasa.gov/Newsroom/view.php?id=42399

[44] http://hypertextbook.com/facts/2000/HannaBerenblit.shtml

[45] http://www.chron.com/commons/readerblogs/atmosphere.html?plckController=Blog&plckBlogPage=BlogViewPost&newspaperUserId=54e0b21f-aaba-475d-87ab-1df5075ce621&plckPostId=Blog%3a54e0b21f-aaba-475d-87ab-1df5075ce621Post%3aa2b394cc-5b5f-47ad-8bb5-c1aec91409ad&plckScript=blogScript&plckElementId=blogDest

[46] http://www.telegraph.co.uk/comment/columnists/christopherbooker/7062667/Pachauri-the-real-story-behind-the-Glaciergate-scandal.html

[47] http://wattsupwiththat.com/2010/01/23/breaking-news-scientist-admits-ipcc-used-fake-data-to-pressure-policy-makers/

[48] http://www.newscientist.com/article/dn18363-debate-heats-up-over-ipcc-melting-glaciers-claim.html?DCMP=OTC-rss&nsref=online-news

[49] Huss, M., M. Funk, and A. Ohmura (2009), Strong Alpine glacier melt in the 1940s due to enhanced solar radiation, *Geophys. Res. Lett.*, 36, L23501, doi:10.1029/2009GL040789.

[50] http://eureferendum.blogspot.com/2010/01/and-now-for-amazongate.html

[51] D. C. Nepstad, A. Veríssimo, A. Alencar, C. Nobre, E. Lima, P. Lefebvre, P. Schlesinger, C. Potter, P. Mountinho, E. Mendoza, M. Cochrane, V. Brooks, Large-scale Impoverishment of Amazonian Forests by Logging and Fire, Nature, 1999, Vol 398, 8 April, pp505.

[52] http://nofrakkingconsensus.blogspot.com/2010/01/more-dodgy-citations-in-nobel-winning.html

[53] http://nofrakkingconsensus.blogspot.com/2010/01/greenpeace-and-nobel-winning-climate_28.html

[54] http://www.telegraph.co.uk/earth/environment/climatechange/7111525/UN-climate-change-panel-based-claims-on-student-dissertation-and-magazine-article.html

[55] http://climatequotes.com/

[56] http://wattsupwiththat.files.wordpress.com/2009/05/surfacestationsreport_spring09.pdf

[57] http://www1.ncdc.noaa.gov/pub/data/uscrn/documentation/program/X030FullDocumentD0.pdf

[58] http://www.theclimateconspiracy.com/files/climategate/FOIA/mail/1257546975.txt

[59] http://chiefio.wordpress.com/

[60] http://www.travel-bolivia.com/bolivia-climate.html#Altiplano%20Region

[61] http://www.theclimateconspiracy.com/files/climategate/FOIA/mail/0942777075.txt

[62] http://www.theclimateconspiracy.com/files/climategate/FOIA/mail/1141398437.txt

[63] http://climateaudit.org/2009/12/10/ipcc-and-the-trick/

[64] http://theclimateconspiracy.com/files/climategate/FOIA/mail/0938031546.txt

[65] http://theclimateconspiracy.com/files/climategate/FOIA/mail/0938031546.txt

[66] http://www.realclimate.org/index.php/archives/2004/12/myths-vs-fact-regarding-the-hockey-stick/

[67] http://theclimateconspiracy.com/files/climategate/FOIA/mail/1047388489.txt

[68] http://theclimateconspiracy.com/files/climategate/FOIA/mail/1047388489.txt

[69] http://theclimateconspiracy.com/files/climategate/FOIA/mail/1047474776.txt

[70] http://theclimateconspiracy.com/files/climategate/FOIA/mail/1051156418.txt

[71] http://theclimateconspiracy.com/files/climategate/FOIA/mail/1051190249.txt

[72] http://theclimateconspiracy.com/files/climategate/FOIA/mail/1051202354.txt

[73] http://theclimateconspiracy.com/files/climategate/FOIA/mail/1054757526.txt

[74] http://theclimateconspiracy.com/files/climategate/FOIA/mail/1057941657.txt

References

[75] http://theclimateconspiracy.com/files/climategate/FOIA/mail/1057941657.txt

[76] http://theclimateconspiracy.com/files/climategate/FOIA/mail/1057941657.txt

[77] http://theclimateconspiracy.com/files/climategate/FOIA/mail/1057941657.txt

[78] http://theclimateconspiracy.com/files/climategate/FOIA/mail/1065125462.txt

[79] http://theclimateconspiracy.com/files/climategate/FOIA/mail/1065125462.txt

[80] http://theclimateconspiracy.com/files/climategate/FOIA/mail/1106322460.txt

[81] http://theclimateconspiracy.com/files/climategate/FOIA/mail/1106322460.txt

[82] http://theclimateconspiracy.com/files/climategate/FOIA/mail/1106322460.txt

[83] http://theclimateconspiracy.com/files/climategate/FOIA/mail/1106322460.txt

[84] http://theclimateconspiracy.com/files/climategate/FOIA/mail/1107454306.txt

[85] http://theclimateconspiracy.com/files/climategate/FOIA/mail/1109021312.txt

[86] http://theclimateconspiracy.com/files/climategate/FOIA/mail/1212063122.txt

[87] http://theclimateconspiracy.com/files/climategate/FOIA/mail/1212063122.txt

[88] http://theclimateconspiracy.com/files/climategate/FOIA/mail/1228330629.txt

[89] http://theclimateconspiracy.com/files/climategate/FOIA/mail/1074277559.txt

[90] http://theclimateconspiracy.com/files/climategate/FOIA/mail/1076359809.txt

[91] http://theclimateconspiracy.com/files/climategate/FOIA/mail/1092167224.txt

[92] http://theclimateconspiracy.com/files/climategate/FOIA/mail/1120676865.txt

[93] http://theclimateconspiracy.com/files/climategate/FOIA/mail/1056478635.txt

[94] http://theclimateconspiracy.com/files/climategate/FOIA/mail/0826209667.txt

[95] http://theclimateconspiracy.com/files/climategate/FOIA/mail/0962818260.txt

[96] http://theclimateconspiracy.com/files/climategate/FOIA/mail/1105019698.txt

[97] http://theclimateconspiracy.com/files/climategate/FOIA/mail/1102687002.txt

[98] http://theclimateconspiracy.com/files/climategate/FOIA/mail/1119957715.txt

[99] http://theclimateconspiracy.com/files/climategate/FOIA/mail/1119957715.txt

[100] http://theclimateconspiracy.com/files/climategate/FOIA/mail/1120593115.txt

[101] http://theclimateconspiracy.com/files/climategate/FOIA/mail/1077829152.txt

Made in the USA
Middletown, DE
18 February 2019